AF341614

MANUFACTURING PROCESS DESIGN AND OPTIMIZATION

MANUFACTURING ENGINEERING AND MATERIALS PROCESSING
A Series of Reference Books and Textbooks

FOUNDING EDITOR

Geoffrey Boothroyd
University of Rhode Island
Kingston, Rhode Island

1. Computers in Manufacturing, *U. Rembold, M. Seth, and J. S. Weinstein*
2. Cold Rolling of Steel, *William L. Roberts*
3. Strengthening of Ceramics: Treatments, Tests, and Design Applications, *Harry P. Kirchner*
4. Metal Forming: The Application of Limit Analysis, *Betzalel Avitzur*
5. Improving Productivity by Classification, Coding, and Data Base Standardization: The Key to Maximizing CAD/CAM and Group Technology, *William F. Hyde*
6. Automatic Assembly, *Geoffrey Boothroyd, Corrado Poli, and Laurence E. Murch*
7. Manufacturing Engineering Processes, *Leo Alting*
8. Modern Ceramic Engineering: Properties, Processing, and Use in Design, *David W. Richerson*
9. Interface Technology for Computer-Controlled Manufacturing Processes, *Ulrich Rembold, Karl Armbruster, and Wolfgang Ülzmann*
10. Hot Rolling of Steel, *William L. Roberts*
11. Adhesives in Manufacturing, *edited by Gerald L. Schneberger*
12. Understanding the Manufacturing Process: Key to Successful CAD/ CAM Implementation, *Joseph Harrington, Jr.*
13. Industrial Materials Science and Engineering, *edited by Lawrence E. Murr*
14. Lubricants and Lubrication in Metalworking Operations, *Elliot S. Nachtman and Serope Kalpakjian*
15. Manufacturing Engineering: An Introduction to the Basic Functions, *John P. Tanner*
16. Computer-Integrated Manufacturing Technology and Systems, *Ulrich Rembold, Christian Blume, and Ruediger Dillman*
17. Connections in Electronic Assemblies, *Anthony J. Bilotta*
18. Automation for Press Feed Operations: Applications and Economics, *Edward Walker*
19. Nontraditional Manufacturing Processes, *Gary F. Benedict*
20. Programmable Controllers for Factory Automation, *David G. Johnson*
21. Printed Circuit Assembly Manufacturing, *Fred W. Kear*
22. Manufacturing High Technology Handbook, *edited by Donatas Tijunelis and Keith E. McKee*
23. Factory Information Systems: Design and Implementation for CIM Management and Control, *John Gaylord*
24. Flat Processing of Steel, *William L. Roberts*

Additional Volumes in Preparation

MANUFACTURING PROCESS DESIGN AND OPTIMIZATION

BY ROBERT F. RHYDER

MARCEL DEKKER, INC.   NEW YORK · BASEL · HONG KONG

FIRST INDIAN REPRINT, 2014

Library of Congress Cataloging-in-Publication Data

Rhyder, Robert F.
   Manufacturing process design and optimization / by Robert F.
Rhyder.
       p.   cm. -- (Manufacturing engineering and materials processing
   : 50)
   Includes index.
   ISBN 978-0-8247-9909-0
   1. Manufacturing process.   2. Production engineering.
I. Title.   II. Series.
TS183.R538 1997
670--dc21

The publisher offers discounts on this book when ordered in bulk quantities. For more information, write to Special Sales/Professional Marketing at the address below.

Copyright © 1997 by MARCEL DEKKER, INC. All Rights Reserved.

Neither this book nor any part may be reproduced or transmitted in any form or by any means, electronic or mechanical, including photocopying, microfilming, and recording, or by any information storage and retrieval system, without permission in writing from the publisher.

MARCEL DEKKER, INC.
270 Madison Avenue, New York, New York 10016
http://www.dekker.com

Printed and bound in India by Bhavish Graphics.

FOR SALE IN SOUTH ASIA ONLY

To my father,
whose mechanical mind
never ceases to amaze me.
And to my daughters,
April and Toni,
the lights of my life.

Preface

This text is an introduction to process optimization techniques. It is designed specifically for practicing engineers and manufacturing supervisory personnel. It presents the basic concepts of process design, problem solving, and process optimization in a straightforward and easy-to-understand manner. I have specifically avoided complex mathematical relationships wherever possible, stressing instead the *concepts of improvement* and *how to use these tools* in the so called "real world."

The first part of the text discusses process design and how specific designs introduce variation into manufactured products. It goes on to present techniques to minimize these variations. These methods would be part of the design stage; however, many can be added after the process has been established.

The second part covers basic problem identification and problem solving methods. These methods involve, among other things, simple techniques for organization and presentation of data that most operations already track, and a few simple techniques of observation. Often, these methods are all that are needed to produce truly dramatic improvements in a manufacturing process and profitability.

Part 3 introduces basic experimental design principles, extremely powerful tools for increasing our understanding of the relationships between the inputs and outputs of manufacturing operations. The text is written to introduce these concepts in a fashion that will enable their effective use, without concentrating on the statistics surrounding their formulation. Specific guidelines for the use of these methods are presented. And although complex statistical analysis of results is purposely not covered, even without the statistical techniques, experimental design principles are still powerful tools.

While there are many examples of actual applications that could have been included in this book, I have limited their use to a few basic ones that demonstrate the concepts being discussed in each chapter. I have purposely avoided complex statistical analysis of the results. All the examples in this book are real, and many of them demonstrate that process improvement efforts do not need to be complex to be effective. Simplicity in design and in the presentation of results is the most effective way to gain the support and involvement of a manufacturing organization as a whole. People support what they understand and merely tolerate what they don't. For the techniques in this book to have a lasting effect on an organization, they must be communicated in a fashion that maximizes understanding and involvement for all levels and functions.

An acquaintance of mine, H. Russell Hastings, once said, "Initiative is doing what needs to be done without being told to do it. In lieu of that, it is doing it after being told only once." By his definition, my initiative with regard to this work was terribly lacking. Having had success implementing and explaining these techniques to coworkers, customers, and suppliers for more than a decade, the feedback I received consistently indicated a need for someone to explain these tools in an easy-to-understand fashion. For years, many individuals urged me to apply my explanations to paper and write this book. The more I traveled and saw American industry in action, the more convinced I was that many of the techniques in this book remain largely misunderstood and underutilized.

My primary focus has been to make the principles in this text understandable to as many people as possible. It would have been very easy to allow the subject matter to become complex, in terms of both math and application. It was quite a bit more difficult to explain these topics in a simple fashion, while ensuring that the approach does not violate the mathematical principles on which these techniques are founded. It is my sincere hope that this book will provide a solid understanding of basic optimization techniques and motivate readers to begin implementing these tools in their manufacturing processes. After the first few successes, I am sure their excitement, interest, and utilization of these techniques will expand rapidly.

No one writes anything in a vacuum. I must thank a few individuals who have helped to make this text easier to follow and comprehend.

First, I want to express my gratitude to my wife Glenda, a quality professional and Certified Quality Engineer, she spent many hours reading drafts of this work and kept it on track. I also wish to express my appreciation to H. Russell Hastings and Michael Slawter, two of the best practicing engineers I know. Their insight and suggestions kept the examples accurate and provided guidance for the expansion of various topics covered in this text.

Robert F. Rhyder

Contents

Part 2: Basic Improvement Techniques

Part 3: Design of Experiments

Part 1

Manufacturing Process Design

1

Process Principles

As manufacturers, many of us tend to view processes as singular entities. How many times have you heard personnel in your organization refer to "the process" or "the line," as if a singular noun adequately described all operations? As we become familiar with the operations of our manufacturing processes, we tend to lump them together and treat them as one continuous machine. While this works fine in casual conversation, this perspective can be deceptive, for we accept processes without examining and truly understanding their characteristics. We often fail to recognize the amount of variability introduced by the operations of our processes. In fact, excessive variability in manufactured products is often the direct result of the process design itself.

Before we begin exploring techniques for process optimization, we shall examine a few basics of process design. As we shall see, certain characteristics are inherent in process design and actually determine the variation in our manufacturing processes. Once we understand these characteristics, we can either eliminate them from our designs, or develop controls to monitor process operations.

Process Definition

For our purposes, a process is defined as follows:

A process is one or more actions on inputs designed to produce an output with one or more specific characteristics.

While many of the concepts which follow could be applied to services as well as products, our terminology will emphasize manufactured products.

A process is comprised of an action or a series of actions taken to transform something (the input) into something else (the output). Graphically, we can represent a process in simplest form by the illustration shown in Figure 1-1.

Figure 1-1: A Simple Process

It certainly looks easy enough, but few processes examined are this simple. A process design examined in Chapter 3 is capable of producing many millions of variations of the same product. This is an amazing amount of variation when we consider that the designers undertook the development of a process to produce a consistent, high quality product.

Normally, a manufacturing process consists of many machines or actions arranged in a definitive sequence to transform a product, or products, with specific input characteristics, into an output with specific but altered characteristics. To understand how processes can become convoluted, we must first examine process structures. After a few process structures have been defined, we will study the effect they have on product variation when they are joined together to form a manufacturing process.

Series Structure

The simplest type of process component is one machine or operation through which all inputs must pass in order to be transformed into outputs. This structure is illustrated in Figure 1-2, below.

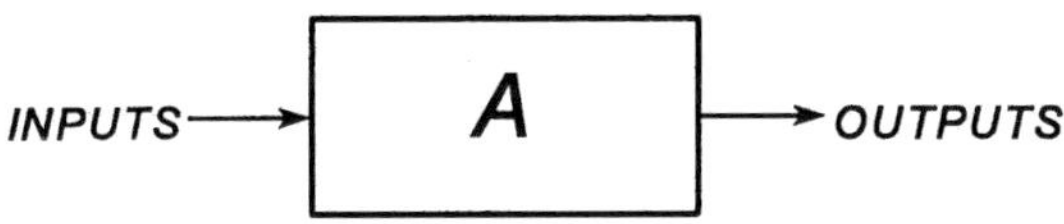

Figure 1-2: A Series Component

The most natural and frequent variation of a simple process is to arrange multiple components in sequence. This is necessary to transform input characteristics into the desired output characteristics, since one machine or operation cannot perform all of the operations needed to produce a product. A series process results whenever we have more than one series component in sequence, as shown in Figure 1-3.

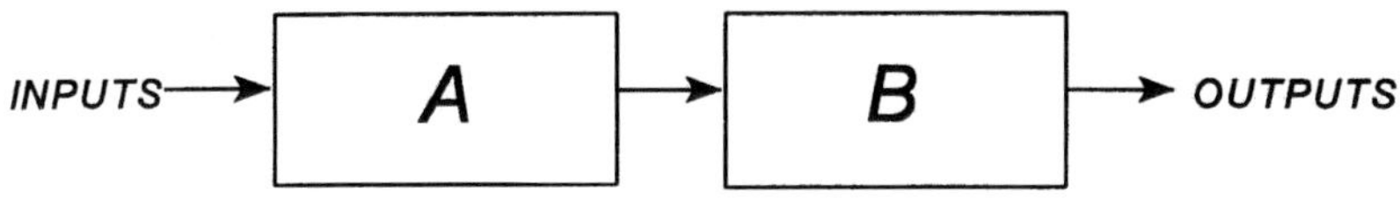

Figure 1-3: A Series Process

In the illustration above, two distinct operations or actions are denoted as A and B. The use of the terms "inputs" and "outputs" recognizes the possibility of multiple characteristics for each. For example, the process may be producing one physical part with multiple measurable output characteristics such as length, width, and height.

Real manufacturing processes are rarely as simple as illustrated in Figure 1-3. It is not unusual for processes to have many operations arranged in sequence to transform the inputs into a product with desired output characteristics. The importance of the figure lies with the realization that the actions A and B represent unique steps in the process. Each represents a possible source of error and, therefore, each has the potential to induce variation in the final product. We shall assume this variation is minimal for now, and concentrate on other sources in the following sections. For some readers, this assumption may not be valid. Individual operations may be the prime source of failure in their manufactured product.

Parallel Structure

The market for a product is often sufficient to support more than one manufacturing line. If we consider the process represented by Figure 1-3 again, we could easily envision a plant where two such manufacturing processes were utilized to meet the market demands for volume. In this case, the process would be represented by the flow diagram in Figure 1-4.

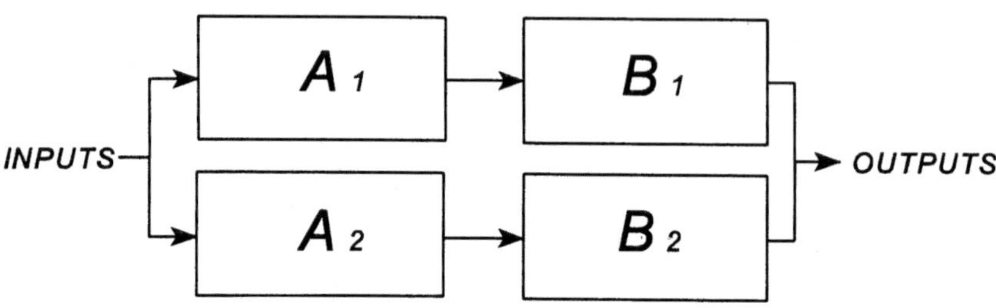

Figure 1-4: A Parallel Process

The subscripts in the figure above represent unique machines and/or operators working to produce product concurrently. The components or machines from the original process are denoted by the subscript "1" after each component letter. The new or added process is denoted by the subscript "2." The complete manufacturing process is comprised of two individual lines or branches.

A branch is defined as two or more process components connected in series.

In this type of arrangement, it is quite normal for each manufacturing line to produce slightly different products. Much effort is placed upon keeping the differences to a minimum. This "minimum" is normally contained within the engineering specifications for the product being manufactured. Despite the example of other countries, such as Japan (where many companies continually strive to achieve the nominal dimension), the current thinking in many U.S. industries dictates the product is acceptable if the measured output characteristics fall within the tolerance band established by the engineering specifications.

Many times, the duplication of all manufacturing process components is unnecessary to meet market demands. In such situations only components which are operating at or near capacity are duplicated to increase production. The result is a parallel component as a portion of the manufacturing process. A simple parallel component is illustrated in Figure 1-5.

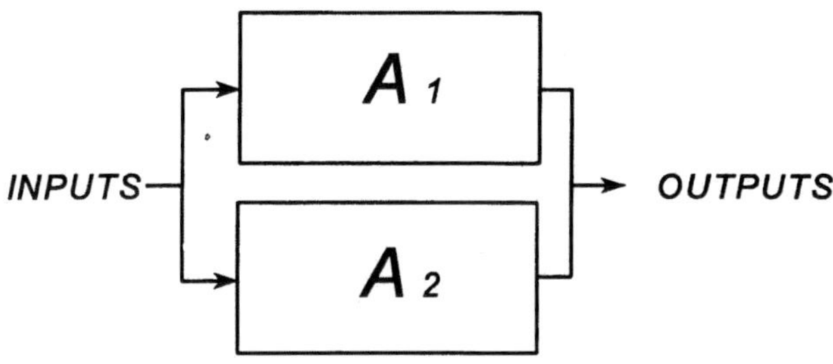

Figure 1-5: A Parallel Component

Notice that outputs from A_1 and A_2 are mixed together immediately after the parts are processed through these operations. This is a key feature in our definition of a parallel component. In the parallel process illustrated in Figure 1-4, the outputs were not mixed together until the end of the process. The distinction between a parallel process, and a process with parallel components will become clear in Chapter 2. As we shall see, a process with one or more parallel components causes more variation in the end product.

Complex Structure

Consider the series process illustrated in Figure 1-6. The efficiency rate for each operation is shown in parentheses. In this hypothetical process, operation B is presently running at 100% of capacity.

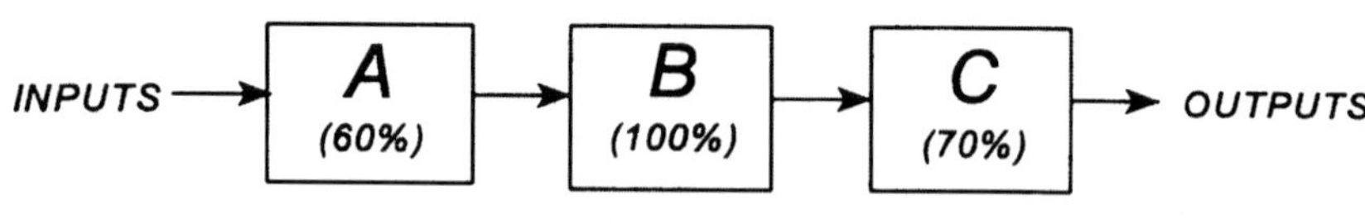

Figure 1-6: Capacity

Obviously, operation B limits the amount of product we can produce. Any requirement to increase production necessitates either improving or duplicating operation B. If we duplicate B to increase production, we get the new flow diagram as shown in Figure 1-7.

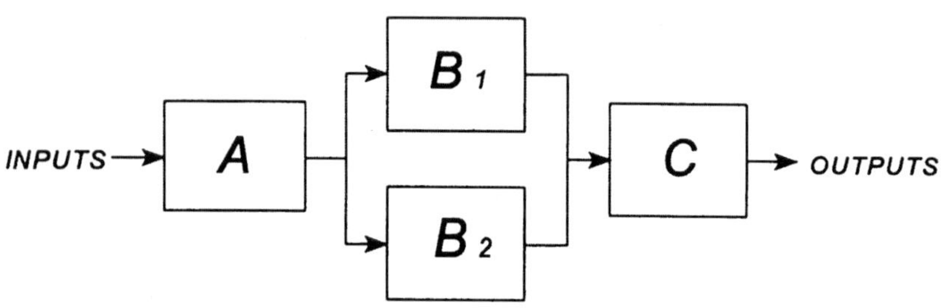

Figure 1-7: A Complex Process

We have replaced the series operation B with a parallel operation consisting of B_1 and B_2. The resulting process is termed "complex" to denote this mixture. Examination of manufacturing processes (especially high speed processes) will reveal many as mixtures of series and parallel components as the process transforms raw materials into finished product. Parallel components will be represented in our process flow diagrams as shown in Figure 1-8.

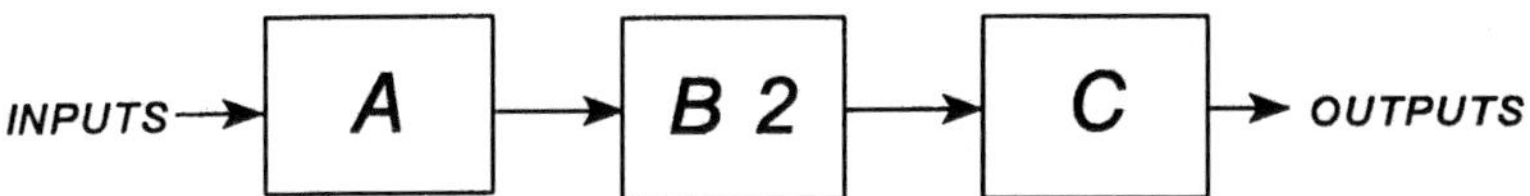

Figure 1-8: Parallel Component Representation

In Chapter 2 we shall examine the variation which may result when complex process designs are used in manufacturing. In Chapter 3, methods will be presented to minimize this variation without adversely affecting the production capacities of the process. We will see how real processes are much more complicated than the simple examples illustrated thus far. Before we begin, let's review a common machine design in use in a variety of manufacturing industries—the rotating table or turret with multiple stations.

Turret Components

In many high speed processes there are operations which are time-dependent. A good example is the act of filling a container in a food processing facility. How quickly the container can be filled is a function of the container volume, product flow rates, temperature, viscosity, and other variables. In this type of operation the process design often includes a rotating table or turret which consists of multiple stations, as illustrated in Figure 1-9. The number of stations (also called pockets or heads) is typically determined by the processing time needed for the operation and the requirement not to slow the speed of other process operations. In Figure 1-9, four stations (S_1–S_4) have been selected to allow the line speed through operations A and C to run at a maximum.

Whether we are discussing filling operations or time-dependent part manufacture, each unit is processed through operation A and enters one of the four available stations in the turret. For instance, the first part made at the beginning of a shift will pass through A and might enter S_1. The operation is performed on the part as it travels partially around the turret. It is then ejected and passes on to C for its final operation. Subsequent parts would enter S_2, S_3, and S_4 in turn. Each station on the turret will process 25% of the product from operation A.

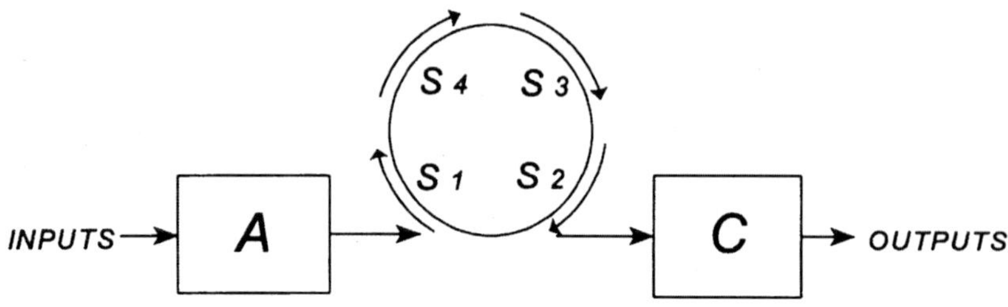

Figure 1-9: Turret Operation

The turret is similar to a parallel component with four stations as shown in Figure 1-10, below. The principal difference is the sequential loading of the rotational component.

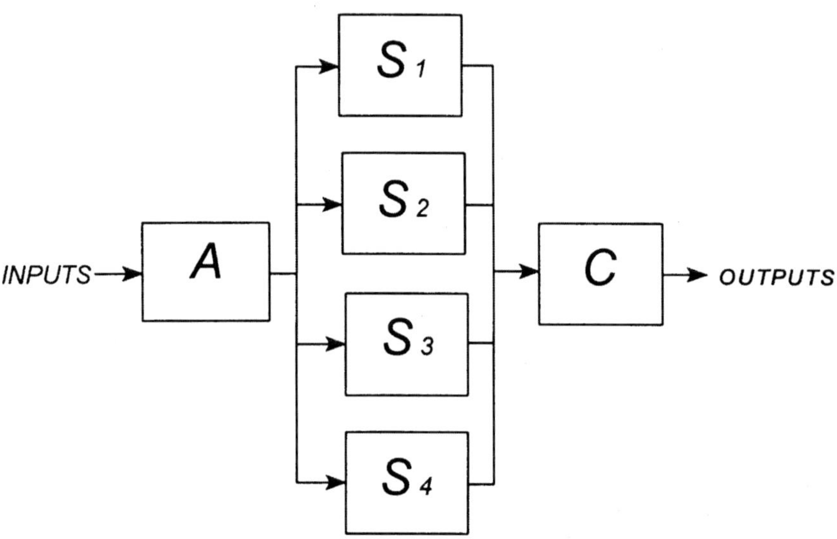

Figure 1-10: Parallel Representation of a Turret

    A turret for time-dependent operations is a reproduction of the operation to maintain the maximum possible effective line speed of a process. As we shall explore later, many manufacturing processes have multiple turrets in a manufacturing process. It is not uncommon to find turrets with fifty or more stations. In general, the higher the flow through capacity of the other operations, the higher the number of stations necessary in a rotational operation.

    A turret should not be confused with a rotary style dial feeder. In some operations, the parts are carried around a table and subjected to a sequence of operations. If the sole purpose of the table is to carry the product from one operation to the next, and the table construction has no effect on the product, then we are dealing with a series process. In this type of design the rotating table is merely acting as a conveyor to transport parts through the operations.

    Generally, whenever the part is carried in a station or a pocket with specified setup procedures, the output characteristics will be determined partly by each station giving us a parallel operation. This is true for time-dependent operations such as fillers, and for sequential operations such as those found in metal stamping facilities. Since these rotational components have unique characteristics from a machine design and manufacturing perspective, it is important to identify them in process flow diagrams. Turret components will be represented in our process flow diagrams as shown in Figure 1-11.

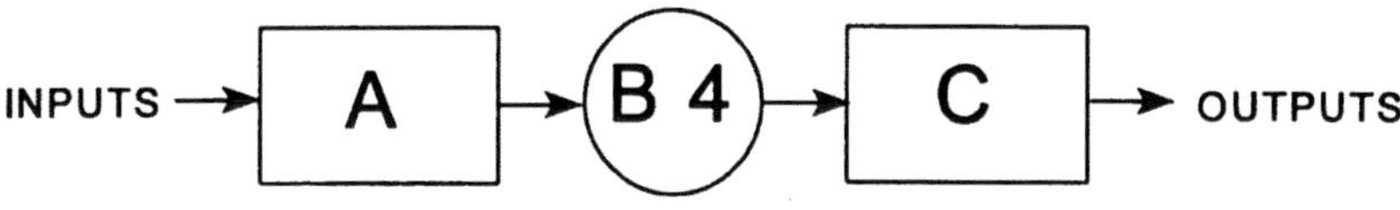

Figure 1-11: Turret Representation

    In this illustration, the turret is process component B, and contains four stations. This representation saves a considerable amount of space when diagraming and analyzing complex processes.

Ovens

Another manufacturing component worthy of review is the processing oven found in many operations. This operation can be either series or parallel, depending upon its use in a process. If we have a single oven used to process one part at a time, we are dealing with a single series component of our process. However, if that same oven is used to process multiple parts simultaneously, we have a parallel component.

There are two basic types of ovens found in manufacturing. The first is the "batch" type, where product is inserted, the door is closed, and product is subjected to a predetermined cycle of temperature. After the completion of the cycle, the product is removed and new product is inserted. The second is the "continuous" type, where product flows through the oven on a chain-link conveyor of one design or another.

Processing ovens do not have truly uniform heat distribution throughout their interior. Typically, much time is spent in the measurement of interior temperatures and the subsequent "balancing" of the oven. While the balancing may be adequate for the manufacture of the product treated in the oven, it should not be taken for granted. If more than one unit of production is treated in the oven at one time, the oven should be represented as a parallel component in the process design. This representation documents the potential for temperature variation across various loading locations in the oven.

If we have a batch type oven, such as in Figure 1-12, the loading of product might be accomplished in a variety of configurations.

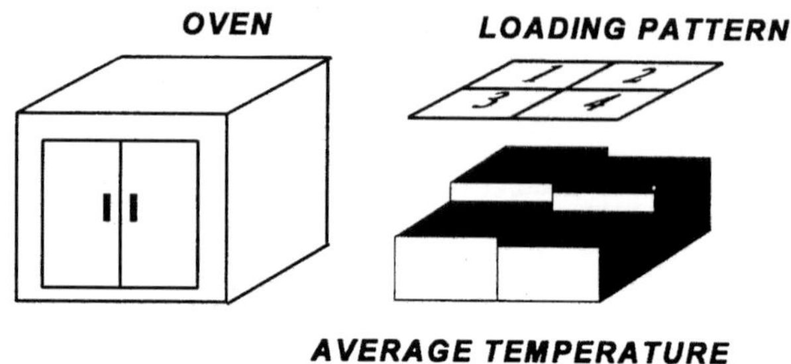

Figure 1-12: Batch Oven Loading and Temperature Distribution

Product might be loaded onto racks inside the oven, or perhaps hung from hooks designed to accommodate the product being manufactured. In either case, representing this type of oven as a parallel process is quite simple. We need only determine the number of locations available in the oven, and include a parallel component representation in our process design with an equal number of stations. For the example shown in Figure 1-12, the component representation would appear as illustrated below.

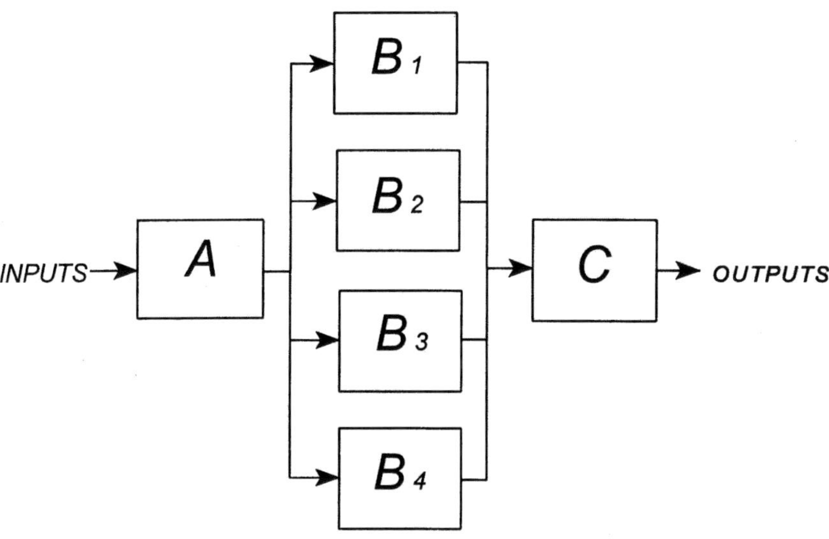

Figure 1-13: Parallel Representation of Batch Oven as Component B

Continuous ovens can be more difficult to quantify than batch ovens for two reasons. First the temperature distribution varies both "down" and "across" the oven, and often fluctuates as a function of the amount of product being loaded. Next, consideration must be given to the method of loading the oven conveyor or belt. Proper representation of this type of component can compensate for both factors.

Loading of the oven can be accomplished in either a controlled or random fashion. If controlled, product will be restricted to specific lanes through the oven's interior. In this case, the oven can easily be represented by a parallel component as illustrated in Figure 1-13, just like the batch oven. This type of loading is shown in the following illustration.

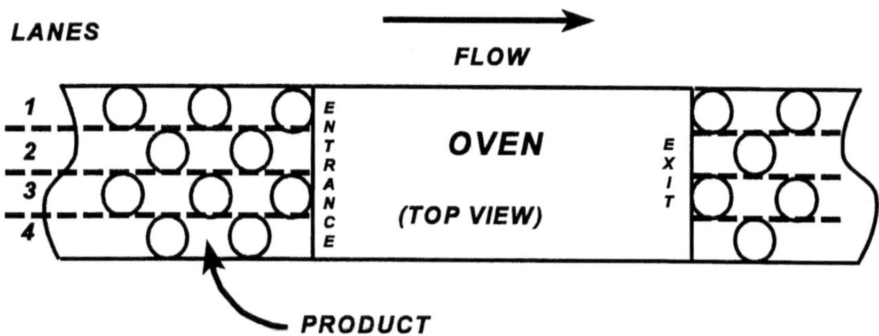

Figure 1-14: Controlled Loading of Continuous Oven

In this example, the product is arranged in four distinct lanes for processing through the oven. The representation of the oven as a parallel component would be the same as shown earlier in Figure 1-13.

In random loading, product is placed on the conveyor without regard to location. This type of loading is illustrated in Figure 1-15. There are no restrictions on product placement within the range of the width of the oven's entrance opening. Technically, this represents an infinite number of potential lanes and, therefore, a parallel component with an infinite number of stations.

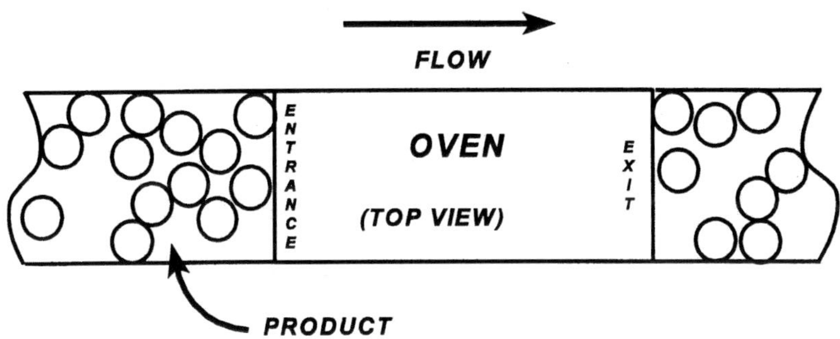

Figure 1-15: Random Loading of Continuous Oven

In practice, one method which might be used to avoid this problem is to divide the width of the oven by the width of the product being processed, and use the resulting quantity (rounded down to the nearest integer) as the "effective" number of lanes. Although not a true representation, this method will allow us to address oven induced variation in finished product. Since the number of potential lanes is theoretically infinite, the "effective" number calculated will always provide for a conservative estimate of variation. The manufacturing process example in the following two chapters will illustrate this concept.

References

1. Bajeria, H. J., Copp, R.P. *Statistical Problem Solving,* Multiface Publishing, Garden City, 1991.

2. Clements, R. B. *Handbook of Statistical Methods in Manufacturing,* Prentice-Hall, Englewood Cliffs, 1991.

3. Juran, J. M. (ed.), Gryna, F. M. (assoc. ed.) *Juran's Quality Control Handbook,* 4th edition, McGraw-Hill, New York, 1988.

2

Process Paths

Now that we've examined basic process structures, the next step is to study how they affect variation in our manufactured products. Often process improvement efforts are stifled by an incomplete knowledge of the underlying process. Through careful examination of process structure, we can gain understanding of the variation inherent in a process design. Once we recognize this variation, we can take steps to improve present processes and utilize better designs for future processes.

Technology has enabled the mass production of virtually all modern products. It normally ensures interchangeability and product function. Throughout industry, many companies have come to rely on technological controls to insure both consistency and volume. Unfortunately, this reliance often causes simple and common-sense designs to be abandoned in favor of more complex technology. That is, many modern designers will develop processes with increasing complexity and rely on electronic controls to overcome shortcomings in process layout and design. Often, they do not even recognize the shortcomings in their process designs. As we shall see in this chapter, it is possible to design processes so complex that no modern controls can ensure consistency of product.

Process Paths

In the examination of manufacturing processes, it is necessary to study the process structure very carefully. Our goal is to determine the exact sequence of operations that must be followed as raw materials are transformed into finished products. Quite simply, the higher the number of possible sequences, the higher amount of variation possible in the finished product. We will define these sequences as follows:

A process path is the unique sequence of operations (performed) on an input.

An input, in the above definition, could mean either raw materials or pre-processed components. With this definition, we can now examine design-induced variation.

Series Paths

A simple series process has one process path, as shown in Figure 2-1.

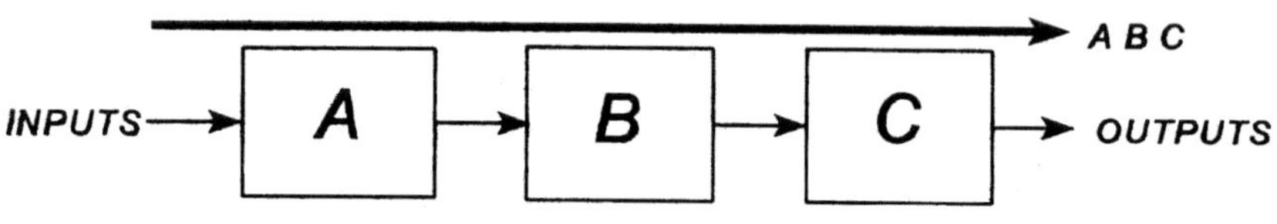

Figure 2-1: A Series Path

The single path is represented by the heavy arrow in the illustration. The arrow follows the precise sequence of operations the inputs must experience in order to become outputs. For the example above, the process path is simply denoted as ABC.

In many processes, rework operations exist at the end of one or more of the manufacturing operations. While stations such as these would add variation to the path, which some inputs could follow during the manufacturing process, we shall assume that rework stations do not exist in our "pure" series process.

Parallel Paths

For parallel processes, the number of paths is determined by the replication of the process operations. A simple parallel process with two branches is illustrated below in Figure 2-2.

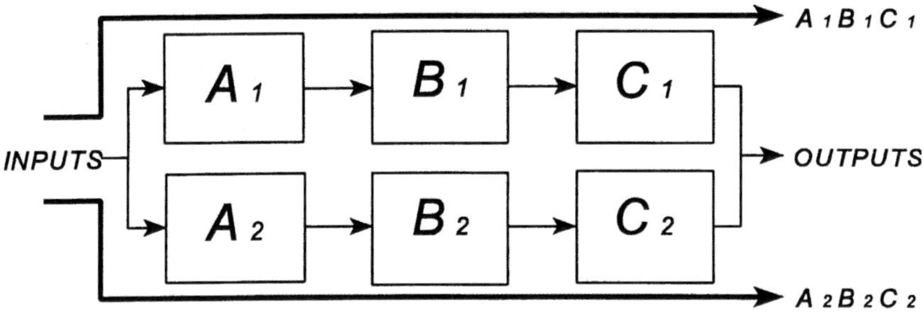

Figure 2-2: Parallel Paths

Once again each process path is shown by a heavy arrow. In this example, two paths are possible for inputs to become outputs. One is $A_1B_1C_1$ and the other is $A_2B_2C_2$. This assumes that product from A_1 is processed only through B_1 and C_1. Similarly, product from operation A_2 may only be processed through B_2 and C_2. (To allow otherwise would constitute a complex process with parallel components.) For example, if the output from A_1 and A_2 were allowed to mix prior to B_1 and B_2, it would be equivalent to Figure 1-7 where operations B_1 and B_2 mixed and became inputs for operation C. This is the distinction between a parallel process (Figure 2-2), and a complex process with parallel components (Figure 1-7). This will become more clear with review of complex process paths in the following section. For now, note that the parallel paths for Figure 2-2 are parallel across all three operations, A, B, and C.

Complex Paths

Determining the number of process paths existing in complex processes is more difficult, although these types of processes are found in most manufacturing environments. The approach to calculating the number of paths requires that we examine each process operation or component, and determine its individual classification. Consider the process shown in Figure 2-3.

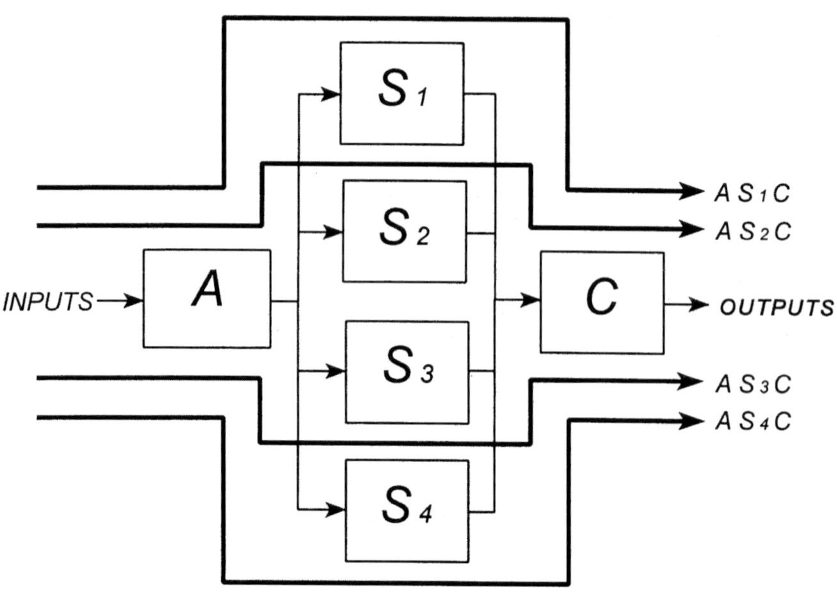

Figure 2-3: Complex Paths

In the illustration above, operation A is a series component connected to a parallel component with stations S_1-S_4. The parallel component is then connected to another series component, C.

From the flow diagram, it is easy to see the four paths product may follow, AS_1C, AS_2C, AS_3C, and AS_4C. We may recognize this illustration as the same one used in Figure 1-9 to depict a process with a four station turret. Adding the turret to the manufacturing process produces four distinct products as output. While the differences between products from each path may be minimal, they do exist. Process designers usually invest a great deal of effort to ensure all of the production from all four stations is within the applicable engineering specification. Unfortunately, we will never find a parallel process where the average output and the variation stemming from each individual station is exactly the same. Even if this could be accomplished with a new installation, it would not last very long. We all know that machine components do not wear at the same rate. All of the stations will begin manufacturing product which is slightly different from the others as machine components begin to wear.

Parallel components also present unique problems from process control and troubleshooting perspectives. We will examine some of the control difficulties when we discuss the application of SPC techniques to parallel operations. As for troubleshooting, the next chapter will provide some insight into methods which may be employed.

Process designs become increasingly complex as series and parallel components are connected in different fashions. They also produce more variation as their complexity increases. The following examples illustrate the basic relationships which are possible.

Example 1: Components in Series

Whenever process designs contain parallel components arranged in series, the number of process paths is multiplicative. If one parallel component with two stations is connected to another with two stations, the number of process paths is 2×2, or 4. This is illustrated in Figure 2-4.

The paths created by this arrangement are A_1B_1, A_2B_1, A_1B_2, and A_2B_2. Each path will produce a slightly different output, determined by the parameters such as setup of individual stations, wear of machine components, and operator.

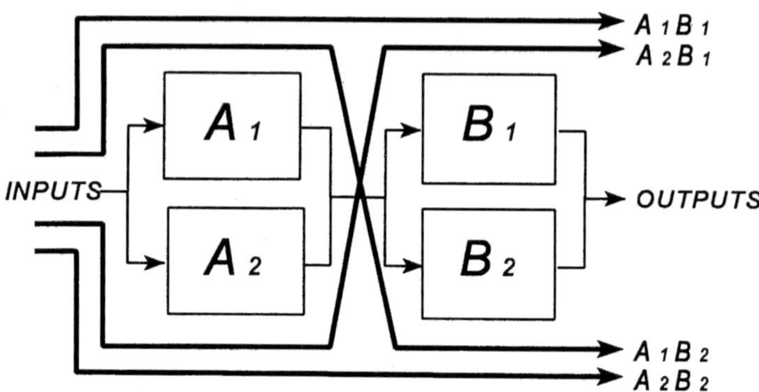

Figure 2-4: Parallel Components in Series

An additional factor which must be considered is the possibility of interactions between stations. For example, A_1 may produce a part with one or more critical dimensional characteristics toward the low end of the manufacturing specification (or merely off target). The setup of B_1 may be such that these low dimensions are of no concern, while the setup of B_2 may cause failure when parts processed from A_1 are run. In this case there is an interaction between stations A_1 and B_2 which causes failure. In complex manufacturing processes, where hundreds or thousands of paths may be possible, such interactions can be exceedingly difficult, if not impossible to identify.

Returning to Example 1, parallel processes connected in series will multiply the number of possible process paths. The number of paths is determined by the number of stations or operations in each individual parallel component in the process. This relationship is shown in Equation (2-1).

$$P = S_{P1} \times S_{P2} \times \ldots \times S_{Pn}$$

where: P = *number of paths*

S_{P1} = *number of stations in parallel component 1*

S_{P2} = *number of stations in parallel component 2*   (2-1)

$\cdot$

$\cdot$

$\cdot$

S_{Pn} = *number of stations in parallel component n*

Consider the process design depicted in Figure 2-5. For this example, there exists a total of 12 process paths. This is determined by the formula in Equation (2-1). We multiply the number of stations in the first parallel component, by the number of stations in the second, by the number of stations in the third, as follows:

$$P = 2 \times 3 \times 2 = 12 \tag{2-2}$$

Notice that series components are not present in the equation. This is because a series component is equivalent to a parallel component with only one station. Including them in the equation would be equivalent to multiplying the parallel component products by one.

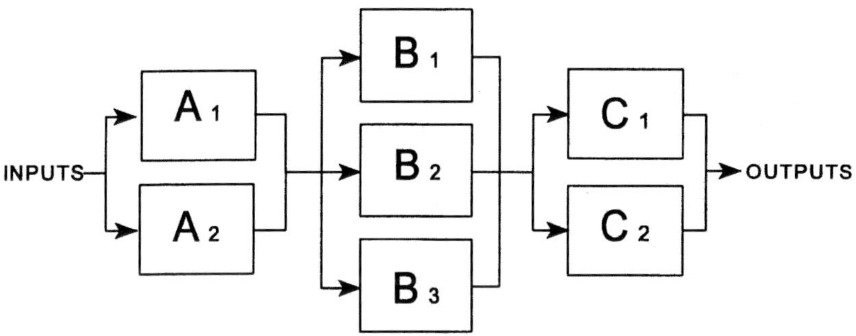

Figure 2-5: Parallel Components in Series

The twelve process paths produced by this series arrangement of parallel components are:

$$A_1B_1C_1$$
$$A_1B_1C_2$$
$$A_1B_2C_1$$
$$A_1B_2C_2$$
$$A_1B_3C_1$$
$$A_1B_3C_2$$
$$A_2B_1C_1$$
$$A_2B_1C_2$$
$$A_2B_2C_1$$
$$A_2B_2C_2$$
$$A_2B_3C_1$$
$$A_2B_3C_2$$

Readers should verify that they are able to visualize each of the twelve possible paths from Figure 2-5. Such visualization is useful in problem solving and process improvement efforts. Take, for example, manufacturing a product with stringent requirements for appearance. If our customers were to complain of a slight scratch or blemish on 50% of the product just received, we could immediately suspect the root cause of the defect occurred in either parallel component A or C, since one defective station in either would affect 50% of our production. Likewise, if the customer complaint was for cosmetic defects in any multiple of 33.3%, we should examine all of the stations in parallel component B. This approach to problem solving will yield positive results in a minimum amount of time. While it is not always effective, it should be the first consideration in root cause investigation. (It is possible that a 50% defective condition is caused by a mixture of stations producing defects. If all product from station B_1 were blemished along with ½ of the product from station B_2 this condition would produce 50% defective product overall.)

In general, any multiple of a parallel component percentage or a process path percentage may provide clues as to the origin of a defective condition. For instance, since Figure 2-5 illustrates a process with 12 paths, any defective condition in multiples of 1/12 (8.3%) may indicate one or more paths are responsible for production of the defect. In other words, an 8% defect rate may indicate one path is responsible, a 16% rate may indicate two paths are responsible, etc.

Example 2: Components in Parallel

Many manufacturing processes contain multiple branches. Within each branch, the components are connected in series. Series connections, as discussed in Example 1, are multiplicative. Whenever multiple branches exist, two or more are connected in parallel. The effect of this parallel arrangement of branches is additive. Consider the process design illustrated in Figure 2-6.

To analyze this design, we must first recognize the process has two branches (just like the parallel process in Figure 1-4). The upper branch consists of stations A_1, B_1, and B_2. The lower branch consists of stations A_2, C_1, and C_2. A visual examination of the upper branch reveals two possible process paths, A_1B_1 and A_1B_2. The lower branch also contains two process paths, A_2C_1 and A_2C_2.

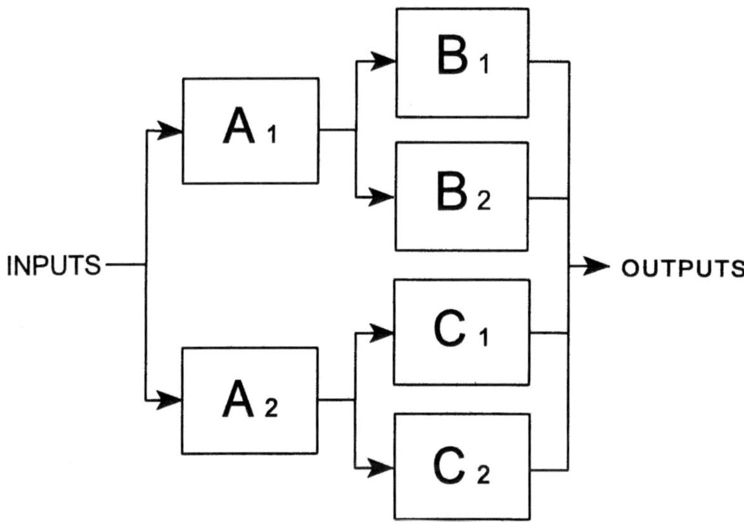

Figure 2-6: Components in Parallel

To determine the total number of paths through this design, we simply add the number of paths from each branch together, according to the following equation:

$$P = B_1 + B_2 + \ldots + B_n$$

where: P = number of process paths

B_1 = number of paths in branch 1

B_2 = number of paths in branch 2   (2-3)

$\cdot$

$\cdot$

$\cdot$

B_n = number of paths in branch n

Figure 2-6 contains a total of four process paths—two from the upper branch, plus two from the lower branch. The key to this analysis is the lack of a connection in the circled region of Figure 2-7.

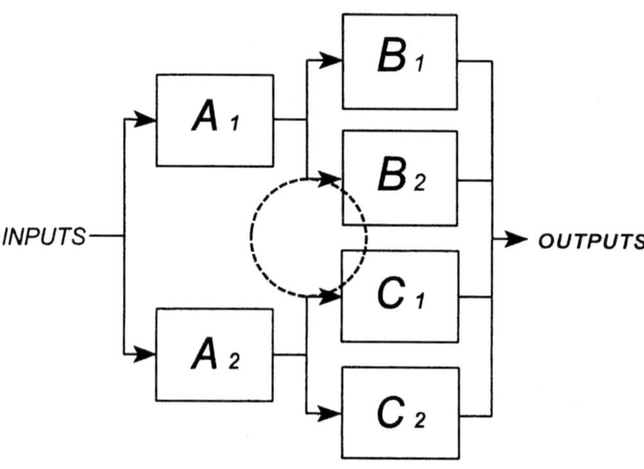

Figure 2-7: Two Branch Parallel Design

Operation A is a parallel component consisting of two stations, A_1 and A_2. In Figure 2-7, however, A_1 is connected in series with the two stations from component B. It is *not* connected to component C. Therefore, A_1 in series with process component B, and A_2 in series with process component C, define two separate branches of this process.

As with all examples thus far, the design illustrated by Figure 2-7 is a simplistic representation for illustrative purposes. The design of the process is balanced, that is, both branches of the process consist of the same type of components arranged as mirror images of one another. Of course, reality is not always this kind, as illustrated in Figure 2-8.

In this arrangement, the upper branch of the process consists of four components—A, B, C, and D. The lower branch, built at a later date to meet increased demands for production quantities, was designed with new technologies, and has a slightly different layout. In the lower branch, component E performs the same function as both components B, and C in the upper branch. Both branches are designed to accomplish the same transformations of inputs into outputs.

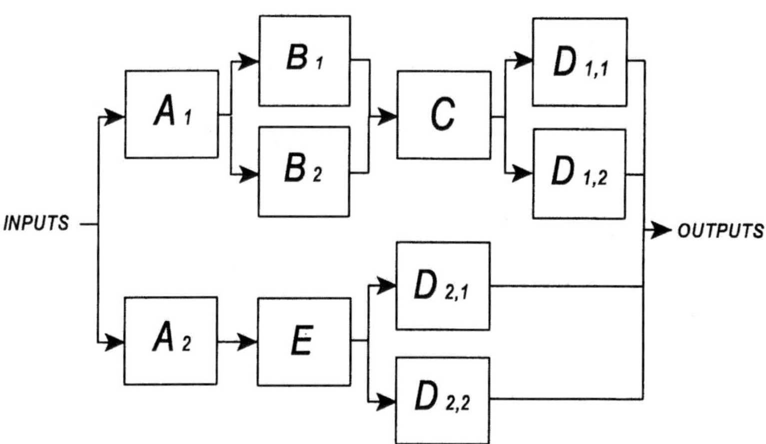

Figure 2-8: Unbalanced Two Branch Process

The notation for component D consists of two subscripts. The upper branch consists of $D_{1,1}$ and $D_{1,2}$. The first subscript indicates the first replicate of the component, while the second subscript indicates the station number. For the lower branch, the first subscript denotes the second replicate of the component while the second indicates the station.

We will assume that ½ of the inputs pass through the upper branch and ½ of the inputs pass through the lower branch. The number of paths in the upper branch is calculated as follows:

$$B_1 = 1 \times 2 \times 1 \times 2$$
$$B_1 = 4$$

(2-4)

The number of paths through the lower branch is:

$$B_2 = 1 \times 1 \times 2$$
$$B_2 = 2$$

(2-5)

The total number of paths is:

$$P = B_1 + B_2$$
$$P = 4 + 2$$
$$P = 6$$

(2-6)

The process is labeled as unbalanced because of the unequal number of paths in each branch. If 50% of production passes through the upper branch, it is equally divided among the four paths present in this branch. This means each path produces 12½% of the total production. For the bottom branch, where only two paths are present, each path will handle 25% of the total production. If we choose to, we can sketch the paths in this arrangement in the form of a tree diagram (Figure 2-9).

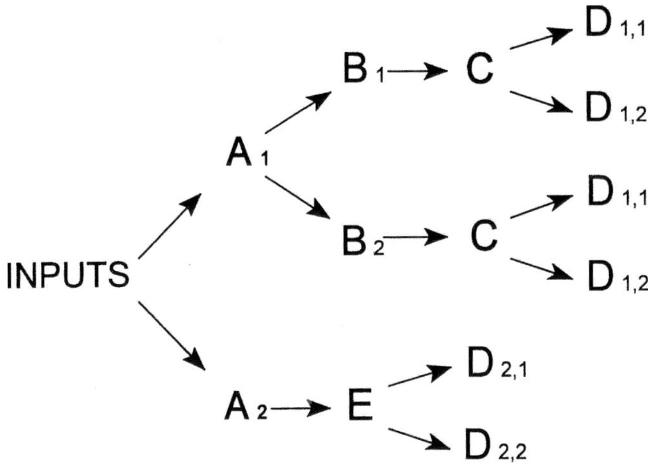

Figure 2-9: Process Paths for Unbalanced Process

A layout of this type can make problem solving more difficult. Suppose we audit production and discover that 25% has light scratches on the exterior. Although this defect may arise from a variety of sources, we should first verify that no single station in the process is responsible for the defect. (This could occur if a setup is performed incorrectly, or if some mechanical subassembly has worn prematurely.) Reviewing the design, we see that 25% of the production will pass through station $D_{2,1}$ or $D_{2,2}$ in the lower branch of the process. We also note 25% will pass through B_1 or B_2 in the upper branch. Because of the unbalanced design, we must check two stations in the upper branch, and two stations from a different component in the lower branch. For highly complex processes, these differences can be very confusing, and will impede our ability to isolate the source of a defect. A careful examination and thorough understanding of the process layout will always enhance problem solving and process improvement efforts.

Example 3: A Complex Manufacturing Process

While product and process identifiers have been necessarily omitted, the complex process illustrated in Figure 2-10 is real. The designers' intent was the production of a very consistent product at an approximate rate of 15,000 units per hour. The entire process, including product and process design, was a collaborative effort between the supplier and the customer from the design phase through first production. Built for almost three million dollars, both the customer and supplier felt confident this design would revolutionize their industry in terms of product safety, ease of use, and appearance.

Unfortunately, after the process was built, the designers found it did not manufacture product in a consistent, reliable manner. As we shall explore, the process design creates a multitude of paths, and therefore, a multitude of unnecessary variation in the product manufactured. In short, the process is so complex that the variation it induces is insurmountable by standard process and tolerance controls.

The process begins with a parallel component containing four stations, denoted as operation A. The process then splits into two balanced branches, denoted by the subscripts 1 and 2. Each branch contains two rotational components, B (with 6 stations) and C (with 35 stations). Both branches are joined and the product is allowed to mix prior to entry into component D, a drying oven whose width is 72 times as wide as the product itself. After curing, the product enters a rotational component with 10 stations for attachment to a subassembly. This is denoted by the letter G in Figure 2-10. After attachment, the product passes through a 100% visual inspection machine and is packed, ready for shipment.

The subassembly consists of plastic parts obtained from two 128-cavity molds, denoted E_1 and E_2. These parts are then subjected to processes F_1 and F_2, each consisting of a 10 station rotational component. According to our definition, the subassembly sequence of operations, E through G, comprise another branch of the overall manufacturing process.

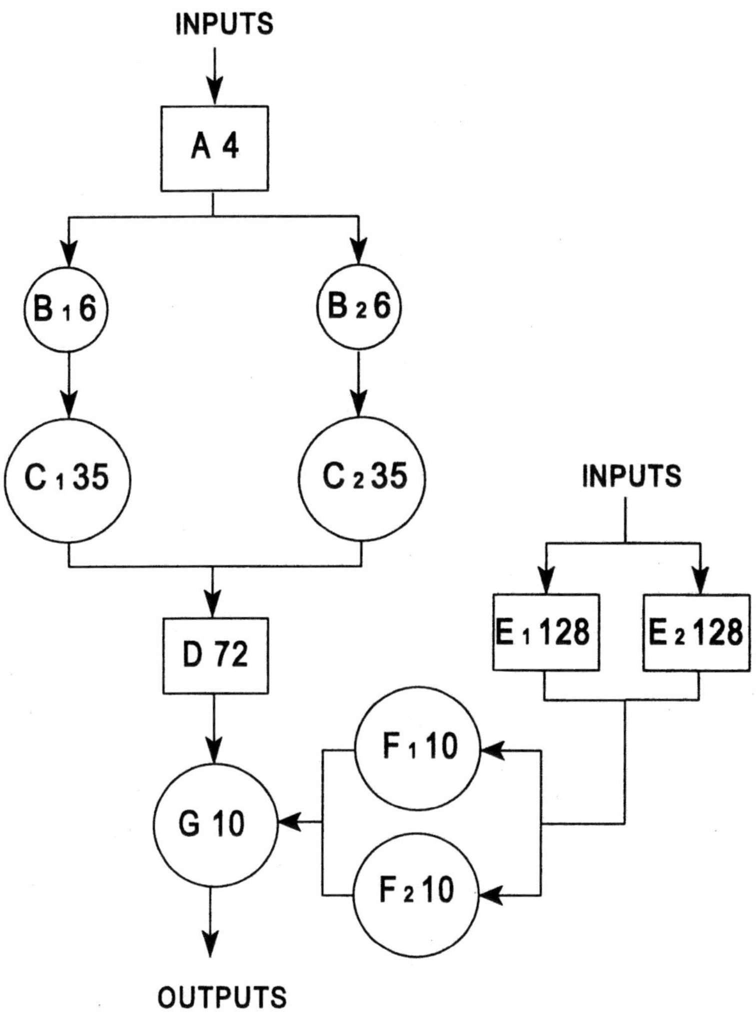

Figure 2-10: Complex Process Example

To calculate the number of process paths present in this design, we shall begin by examining individual branches. Since the branch through components B and C denoted by the subscript 1 is identical in design to the branch denoted by the subscript 2, we need only to calculate the number of paths one time. Each path contains two operations, a rotational component B, with 6 stations, and a rotational component C, with 35 stations. The number of process paths in each branch is:

$$P_{B_1 C_1} = P_{B_2 C_2} = 6 \times 35$$
$$P_{B_1 C_1} = P_{B_2 C_2} = 210 \tag{2-7}$$

To calculate the total number of paths through both branches, we add them together as shown in Equation 2-8. For our present example, the total number of paths through both branches of components B and C is:

$$P_{BC} = P_{B_1 C_1} + P_{B_2 C_2}$$
$$P_{BC} = 210 + 210 \tag{2-8}$$
$$P_{BC} = 420$$

For the branch containing the plastic subassembly, we note there are two 128-cavity molds connected in parallel. The number of process paths through components E_1 and E_2 is simply:

$$P_E = E_1 + E_2$$
$$P_E = 128 + 128 \tag{2-9}$$
$$P_E = 256$$

For components F_1 and F_2, the number of process paths is:

$$P_F = F_1 + F_2$$
$$P_F = 10 + 10 \tag{2-10}$$
$$P_F = 20$$

Recognizing that components E and F are connected in series, we know the total number of process paths in the subassembly branch will be multiplicative. This was illustrated earlier in Equation 2-1. For the subassembly branch, the number of process paths is:

$$P_{EF} = P_E \times P_F$$
$$P_{EF} = 256 \times 20 \tag{2-11}$$
$$P_{EF} = 5120$$

We may now redraw the process flow diagram, and replace the branches by the number of paths they contain. This representation of the process will greatly simplify our diagram, and the subsequent calculations of process paths. This "intermediate" process representation is illustrated in Figure 2-11.

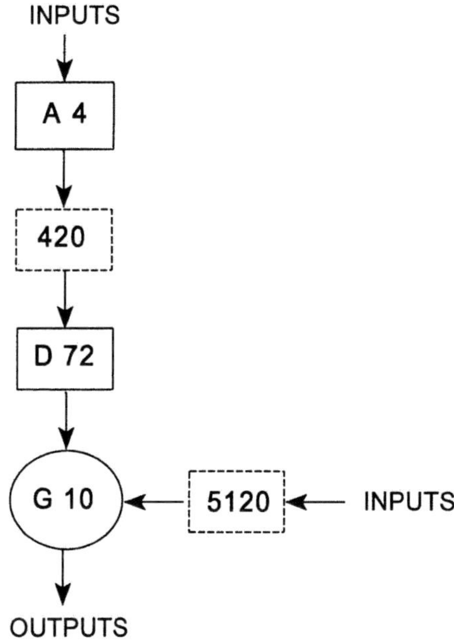

Figure 2-11: Complex Process Example—Intermediate

We now continue the calculation of process paths. Component A has 4 paths connected in series with both branches of components B and C, which is then connected in series to the curing oven, component D. The number of paths is found by multiplying, as shown below:

$$P_{ABCD} = P_A \times P_{BC} \times P_D$$
$$P_{ABCD} = 4 \times 420 \times 72 \qquad\qquad (2\text{-}12)$$
$$P_{ABCD} = 120,960$$

We may once again produce an intermediate flow diagram to illustrate the process paths calculated thus far. This is shown below.

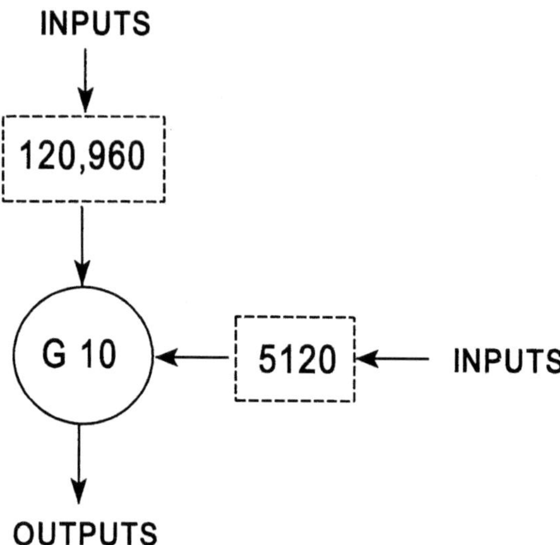

Figure 2-12: Complex Process Example

We now have two distinct input branches feeding the assembly machine, component G. In our earlier examples, we stated that parallel branches were added together to determine the effect they have in producing multiple paths. However, two paths in this example are not parallel. Each branch contains different components of the final product. The subassemblies created by components E and F (the 5120 branch) are loaded into the assembly machine G at random. The subassemblies from components A thru D (the 120,960 branch) are then loaded at random into the assembly machine, once the EF subassembly is in place. To finish our calculation, we must multiply the number of paths in each branch and the number of stations on the assembly machine together. The total number of process paths possible for the complex process example is:

$$P_{A \to G} = P_{ABCD} \times P_{EF} \times P_{G}$$
$$P_{A \to G} = 120{,}960 \times 5120 \times 10 \qquad (2\text{-}13)$$
$$P_{A \to G} = 6{,}193{,}152{,}000$$

Almost 6.2 *billion* ways to manufacture a product. This is an incredible number of process paths. Recall the output of the process was targeted at 15,000 units per hour, or approximately 100,000 units per shift. If we assume the process runs 3 shifts per day, and 250 days per year, this manufacturing process could not possibly produce all of the possible variations in product caused by unique process paths in less than 82.5 years. This assumes no downtime, no wear and replacement of machine components (if components were replaced, that would generate new, unique paths), no variation in raw materials, no variation in the processing environment, and no variation due to operators and production shift. It also assumes the product will still be viable in the marketplace 82.5 years from now. Once we consider all of the above, it becomes obvious that the designers of this process will never see the full range of variation in production due to the multitude of process paths. Imagine, if you can, being responsible for returned product analysis, root cause determination and corrective action for this process.

We stated earlier that once up and running, the line did not meet the designers' expectations. Perhaps now we can see why it did not. In spite of the most modern equipment, and controls, the line did not function as the designers thought it would. Fortunately, there are ways to correct design problems such as these. In the next chapter, we shall examine techniques to minimize variation induced by process designs.

References

1. Bajeria, H. J., Copp, R.P. *Statistical Problem Solving,* Multiface
    Publishing, Garden City, 1991.

2. Clements, R. B. *Handbook of Statistical Methods in Manufacturing,*
    Prentice-Hall, Englewood Cliffs, 1991.

3. Juran, J. M. (ed.), Gryna, F. M. (assoc. ed.) *Juran's Quality Control
    Handbook,* 4th edition, McGraw-Hill, New York, 1988.

3

Optimizing Process Designs

The complex process studied in Chapter 2 illustrates how processes can induce variation into a manufactured product. Even though they were not shown in our process flow diagram, the process included three automated vision inspection systems. One was placed at the end of the line, and two were placed approximately at the midpoint of the line. The designers' intent was to reject nonconforming product early, before adding value. The existence of these devices was often perceived as a demonstration of commitment to quality. It is probably more accurate to assume the designers anticipated difficulties early in the design process. The process was designed for high volume production and the inspection devices were included because the process was expected to go awry.

In this chapter we shall examine a few techniques to improve the design of manufacturing processes. We will revisit the complex process from Chapter 2 to determine how variation from process paths can be reduced. Wherever possible, the cost of the improvements will be kept to a minimum and line speed of the process will not be hindered. Problem solving and process improvement efforts will become much simpler, and faster. Ultimately, the manufactured product will be more consistent.

Isolation

The simplest way to reduce process paths, and therefore potential variation, is to isolate product feeding parallel branches and the branches themselves. For example, consider the process diagram in Figure 3-1:

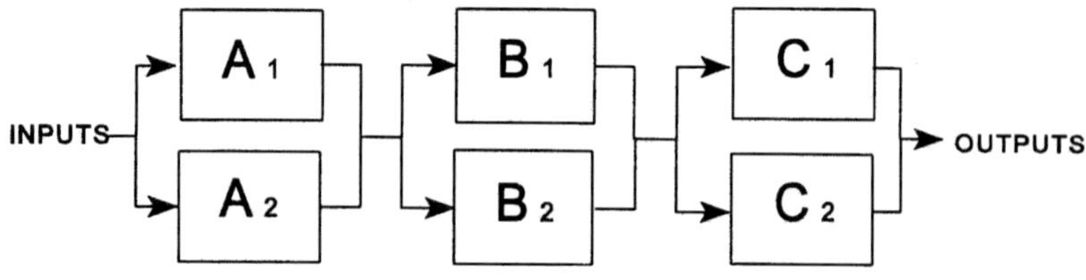

Figure 3-1: Parallel Components in Series

We have 3 parallel components connected in series with one another. Each component contains two stations. The number of paths is:

$$P_{ABC} = 2 \times 2 \times 2$$
$$P_{ABC} = 8$$

(3-1)

If we isolate the component stations, and direct their output the new process flow diagram could appear as follows:

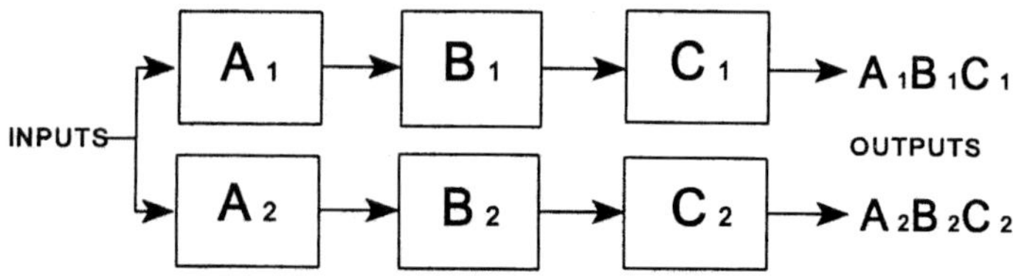

Figure 3-2: Isolated Parallel Components

   With the output from each station isolated and controlled, there are only two possible paths, $A_1B_1C_1$ and $A_2B_2C_2$. This simple act reduces the number of process paths from eight to two. The potential variation from process paths has been reduced 75%.

   Depending upon the process, the mechanism to accomplish isolation may be as simple as a new set of standard operating procedures and operator training. In other cases, a simple physical barrier may be sufficient. In some cases it could be quite a bit more complicated, such as trying to isolate all of the outputs from a 128-cavity plastic mold.

   We can apply this technique to the complex process evaluated in Chapter 2. To accomplish this, let's consider individual branches of the process. We determined when component A was connected in series to the $B_1 C_1$ and the B_2C_2 branches, and the curing oven D, we ended up with 120,960 process paths. This portion of the process is repeated below.

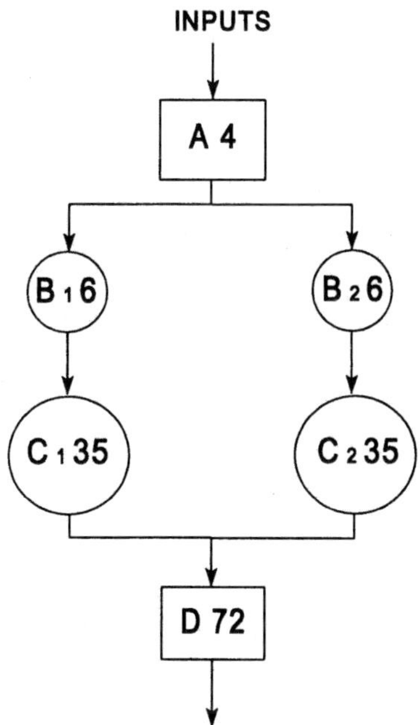

Figure 3-3: Complex Process A → D

The branches for components B_1C_1 and B_2C_2 are already isolated from one another. We need to isolate ½ of component A's four stations and direct them to B_1C_1, then direct the other ½ to B_2C_2. We also need to control the flow of product into the curing oven, component D. This is best accomplished by placing a barrier or diverter on the infeed to the oven. It would certainly be less expensive than placing a barrier inside the oven, and it will not affect the airflow and temperature gradient. With these controls in place, the process flow for components A → D now appears as shown in Figure 3-4.

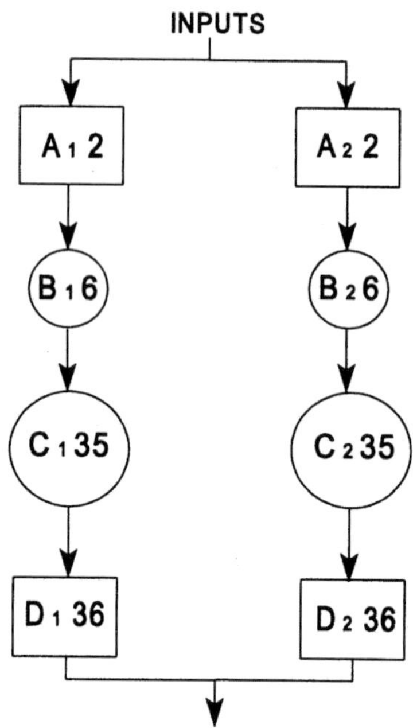

Figure 3-4: Complex Process A → D, Isolated

Isolation reduces the number of possible process paths for this portion of the process from 120,966 to 30,240. This is a 75% reduction in potential paths. Essentially, all we had to do was control the product flow out of component A and add a diverter prior to the entrance of component D, the curing oven. The calculation of paths is shown in Equation 3-2.

$$P_{ABCD} = (2 \times 6 \times 35 \times 36) + (2 \times 6 \times 35 \times 36)$$
$$P_{ABCD} = 15{,}120 + 15{,}120 \tag{3-2}$$
$$P_{ABCD} = 30{,}240$$

Now we shall examine the other input branch of the complex process from Chapter 2, consisting of components E and F. When we calculated the number of paths for this branch, we determined that 5,120 were possible. This portion of the process is illustrated in Figure 3-5.

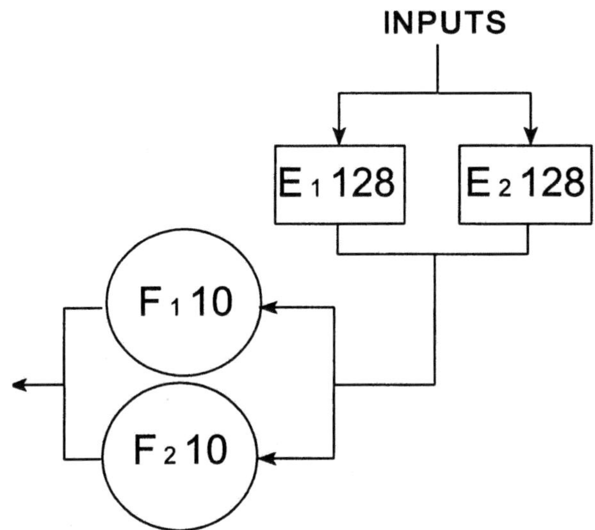

Figure 3-5: Complex Process E → F

As indicated earlier, it could be costly to isolate each of the individual mold cavities on both of the 128-cavity molds. We shall not attempt to do this now, rather, we will merely isolate each mold from the other. We shall follow the same procedure with component F. Isolating station 1 from station 2 will lead to a reduction in the number of process paths without adding any cost to the manufacturing process. The isolated components are illustrated in Figure 3-6.

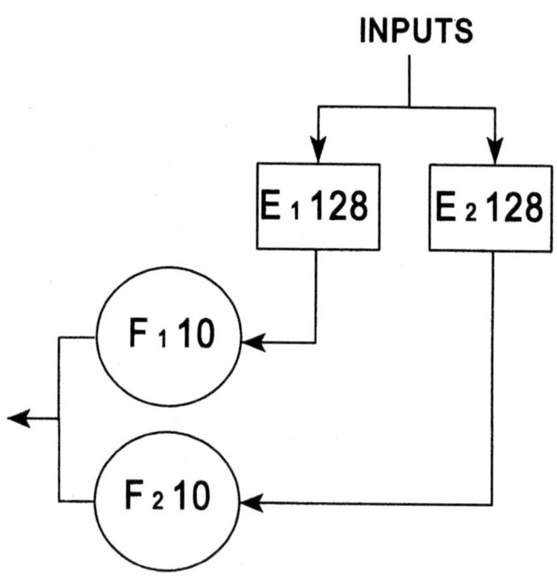

Figure 3-6: Complex Process E → F, Isolated

Once again, isolation has been accomplished with minimal cost and effort. In this example, it is only necessary to insure the product from each mold is kept separate and used as an input to only one of the F components, either F_1 or F_2. The number of process paths through E and F is now:

$$P_{EF} = (128 \times 10) + (128 \times 10)$$
$$P_{EF} = 1,280 + 1,280$$
$$P_{EF} = 2,560$$

(3-3)

The original number of paths through these components was 5,280. The simple act of directing output from each mold of component E has decreased the number of process paths by 50%. If we now consider the entire process, we can determine the effect isolation would have on the almost 6.2 billion process paths we calculated in Chapter 2. Figure 3-7 shows the process with the isolation paths we have calculated.

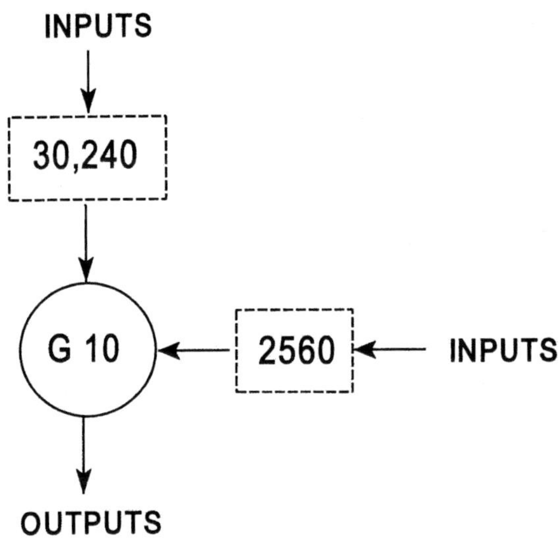

Figure 3-7: Complex Process Example, A → F Isolated

The number of process paths calculated for the entire process with isolation is approximately 774 million. This is shown in Equation 3-4.

$$P_{A \to G} = 30{,}240 \times 2{,}560 \times 10$$
$$P_{A \to G} = 774{,}144{,}000$$

(3-4)

While 774 million process paths is far from a desirable condition for process control and product integrity, it does represent an 87.5% reduction from the 6.2 billion paths calculated earlier. If we calculate the minimum length of time the process would have to run to produce these path variations as we did in Chapter 2, we will find a minimum time requirement of 10.3 years.

Fortunately, isolation is not the only technique we can use to minimize the number of paths defined in process designs. The techniques we are about to discuss cost more than isolation, but would be essential in a design such as the one we have been studying. This is especially important if we are dealing with a product which affects safety or human life.

Multiples and Synchronization

We have seen that components connected in series cause the number of process paths to be multiplicative. That is, if we have a component with 2 stations in series with a 5 station component, we generate 2×5 or 10 process paths. With this multiplicity in mind, consider the initial branches of the complex process again.

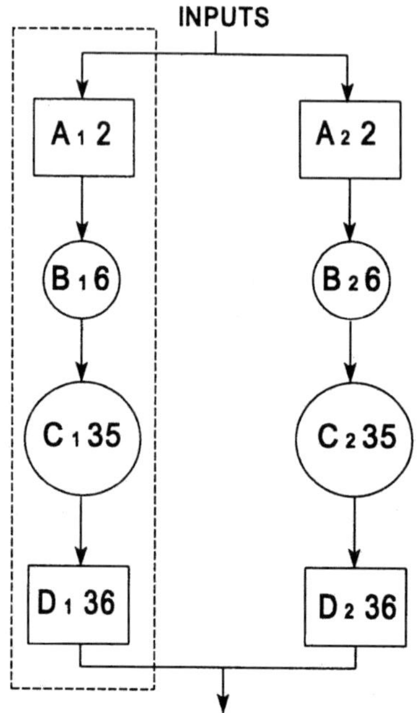

Figure 3-8: B_1C_1 Branch

The branch containing components B_1C_1 has been enclosed by dotted lines. Since both branches are identical, we shall limit our present discussion to the B_1C_1 branch, and duplicate the results on the B_2C_2 branch.

Component A_12 consists of two stations isolated from parallel component A4, which contains 4 stations. It is connected in series with component B_16, a 6 station rotational device. These two components, as illustrated, produce 2×6 or 12 unique process paths. If we denote the two stations from component A_12 as $A_{1,1}$ and $A_{1,2}$, and denote the six stations from component B_16 as $B_{1,1}$ to $B_{1,6}$ respectively, each of the twelve possible paths can be shown in the diagram below.

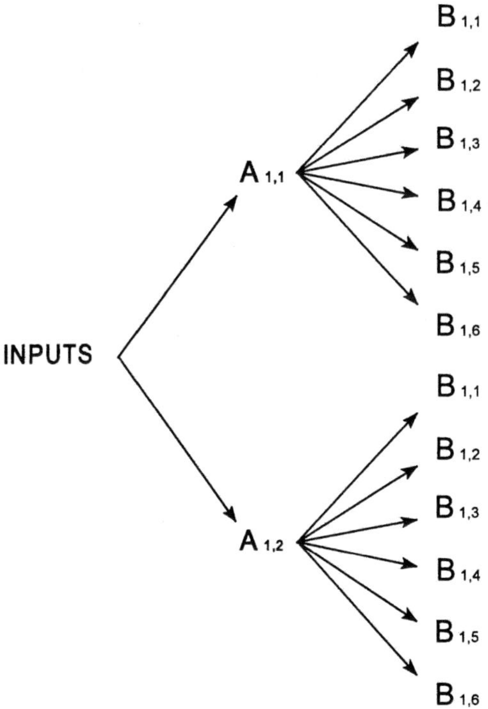

Figure 3-9: Process Paths for Components A and B

The diagram indicates that station $A_{1,1}$ and station $A_{1,2}$ both randomly feed component $B_1 6$. That is, product manufactured through $A_{1,1}$ can pass through any of the six stations of $B_1 6$. The same statement is true for product manufactured through $A_{1,2}$. What would happen if we controlled and directed the flow of product out of component $A_1 2$? This is simply an expansion of what we have already accomplished by diverting two of A's four stations to feed the $B_1 6$ branch, and diverting the other two stations to feed the $B_2 6$ branch. This action goes beyond the simple act of isolating and provides synchronization of the components. As we shall see, it is also more difficult, and usually more costly.

Notice that the number of stations in the rotating component $B_1 6$ happens to be a multiple of the previously isolated component $A_1 2$. Since the 6 stations in $B_1 6$ are a multiple of the 2 stations from $A_1 2$, we can devise methods to synchronize these two components. Our goal is to control the output of $A_1 2$, such that all product passing through station $A_{1,1}$ is fed into specific stations of rotating component $B_1 6$. This might be accomplished simply by isolating the output of $A_1 2$ into two chutes, controlling their entry into component $B_1 6$ with an alternating gate, and synchronizing the gate switching mechanism with the rotation of component $B_1 6$. With this system, product from $A_{1,1}$ can be directed in station $B_{1,1}$ and each second station thereafter. Likewise, product from station $A_{1,2}$ can be directed in $B_{1,2}$ and each second station thereafter. The diagram in Figure 3-10 illustrates the process paths resulting from this setup.

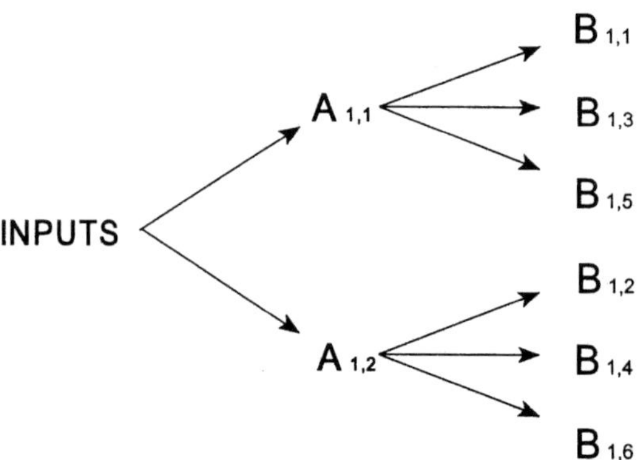

Figure 3-10: Components A and B Synchronized

We have now reduced the 12 process paths to 6 paths by noting the stations in $B_1 6$ were a multiple of A 2, controlling product flow, and synchronizing the two components. Under these conditions, the equation we developed earlier for multiplying the number of stations in $A_1 2$ by the number of stations in $B_1 6$ no longer holds. *The number of process paths for components which are synchronized multiples is simply equal to the number of stations in the component containing the most stations.* Obviously, our process flow diagrams need some sort of notation to indicate when this system structure is in place. For our purposes, we shall use notation as shown in Figure 3-11.

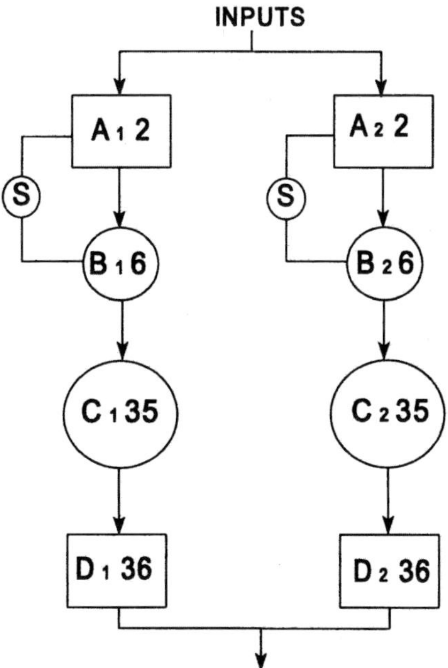

Figure 3-11: Representation of Synchronized Components

The symbols in Figure 3-11 indicate the two operations are synchronized. There is no indication of whether the devices used to accomplish this are based upon a feedback signal from B or from A. This is not important in our present efforts toward increased understanding of process designs, and the effect they may have on path induced variation. It is sufficient to note when two or more components are synchronized and be aware that the number of paths generated by the structure is equal to the number of stations in the "largest" component.

Having established a process structure with 6 paths through components $A_1 2$ and $B_1 6$, we now direct our attention to component $C_1 35$. Obviously, the 35 stations in this component are not a multiple of the 6 we have just established. Let us examine what might have happened if the process designers had attempted to minimize the number of potential paths early in their design process. Component C probably could not be designed with 30 stations without reducing production capacities, however, it certainly could have been constructed with 36 stations. While this would have added mechanical elements to the rotational component C, it would also reduce complexity by providing symmetry in the design, reduce the required rotational velocity of the component, and increase the life of turret components such as bearings, etc. With 36 stations in component C, the process diagram would now appear as illustrated in Figure 3-12.

Since the number of stations in component $C_1 36$ is now a multiple of the number in $B_1 6$, we may reduce the number of potential process paths by synchronizing the two operations. The synchronization may be accomplished mechanically, electronically, or simply by insuring the output of $B_1 6$ is kept sequential, and fed into $C_1 36$ with $B_{1,1}$ feeding $C_{1,1}$. Synchronization of components can be as simple as insuring $B_{1,1}$ feeds $C_{1,1}$ at each line startup (provided there are no inspection/rejection stations between the two components). Startup refers to *every* startup, not merely the commencement of operations at the beginning of each shift. This requires operating procedures and mandates the training of line operators and maintenance personnel to maintain the synchronization after each line stoppage.

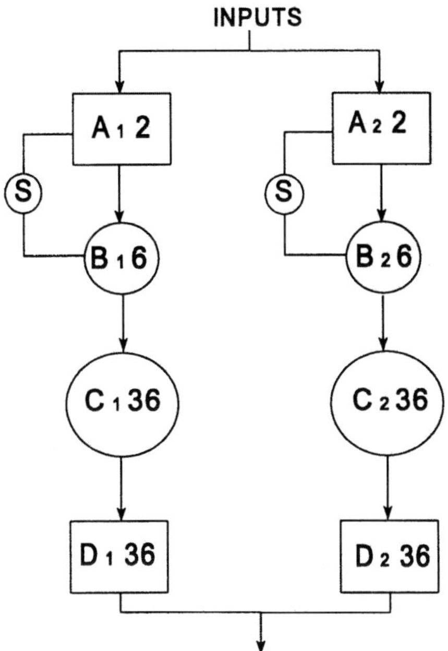

Figure 3-12: Component C with 36 Stations

Next, notice that output from $C_1 36$ can be controlled and synchronized with $D_1 36$. We now have a process design where components $A_1 2$ through $D_1 36$ are all multiples and synchronized. Thus, we can insure the total number of process paths for the branch $A_1 2 \rightarrow D_1 36$ is 36, and the total number of process paths for branch $A_2 2 \rightarrow D_2 36$ is 36. Under these conditions, the total number of process paths through components A, B, C, and D is:

$$P_{ABCD} = P_{A_1 2 \rightarrow D_1 36} + P_{A_2 2 \rightarrow D_2 36}$$
$$P_{ABCD} = 36 + 36 \tag{3-5}$$
$$P_{ABCD} = 72$$

Recall from Chapter 2, that the real manufacturing process produces 120,960 process paths through components A, B, C, and D (this is summarized in Figure 2-12). By using the isolation technique, designing multiples into the process and synchronizing the components, we have reduced the number of process paths 99.94% to a mere 72. The new design does not sacrifice production speed, and certainly improves productivity by reducing potential variation and scrap. The flow diagram for the synchronized process is shown in Figure 3-13.

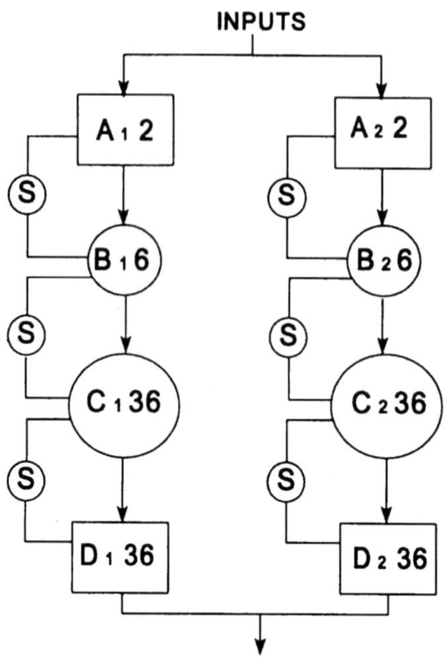

Figure 3-13: Components A → D Synchronized

Having reduced potential variation in the ABCD branches of the process, we now turn our attention to components E and F. Recall from earlier discussion that components E_1 and E_2 are both 128-cavity plastic injection molds. F_1 and F_2 are both 10 station rotational components to process the plastic parts. Earlier we isolated the E_1 from E_2 and F_1 from F_2. This action reduced the number of potential process paths from 5,120 to 2,560. Based upon our present discussions about synchronization, we can easily see that the minimum number of paths possible through these two components is going to be equal to $128 + 128$ or 256. This condition will be achieved if components E_1 and E_2 are multiples of F_1 and F_2 respectively, and they are synchronized.

For components F_1 and F_2, since 128 is not divisible by 10 without a remainder, we must choose between 8 stations and 16 stations. We cannot choose 8 stations unless we can increase the rotational velocity of components F_1 and F_2. Without the increase, 8 stations would slow the line speed and adversely affect productivity. For the sake of this exercise, we will choose to increase the number of stations from 10 to 16 in both of these process components. This yields the process flow diagram for components E and F shown in Figure 3-14.

If we commit to controlling and directing the output from each of the 128-cavity molds, we can ensure 128 paths through $E_1 \rightarrow F_1$ and 128 paths through $E_2 \rightarrow F_2$. (This may sound like an astronomical task, however, in many situations it is possible without prohibitive expense. For example, we can obtain the same improvement by merely dividing the 128 cavities into groups of 16 and directing their flow into component F.)

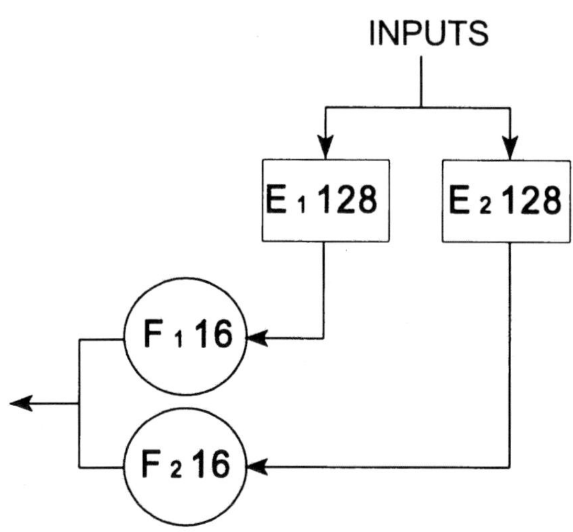

Figure 3-14: Component F with 16 Stations

The number of paths through synchronized components E and F is:

$$P_{EF} = 128 + 128$$
$$P_{EF} = 256$$

(3-6)

Recall from Equation 3-3 that the number of paths in the isolated EF branch was 2,560. If we design components E and F as multiples and synchronize their operation, the number of process paths can be reduced 90% from the simple isolated version. (The original process design produced 5,280 paths through these two components.)

The only component remaining is rotational component G, with 10 stations. Since P_{ABCD} equals 72, and P_{EF} equals 256, G should be designed in a multiple of 8. Since the original design specified 10 stations, and we do not want to slow the process, the new design should have at least 16 stations. This design for component G synchronized with components E and F is shown in Figure 3-15.

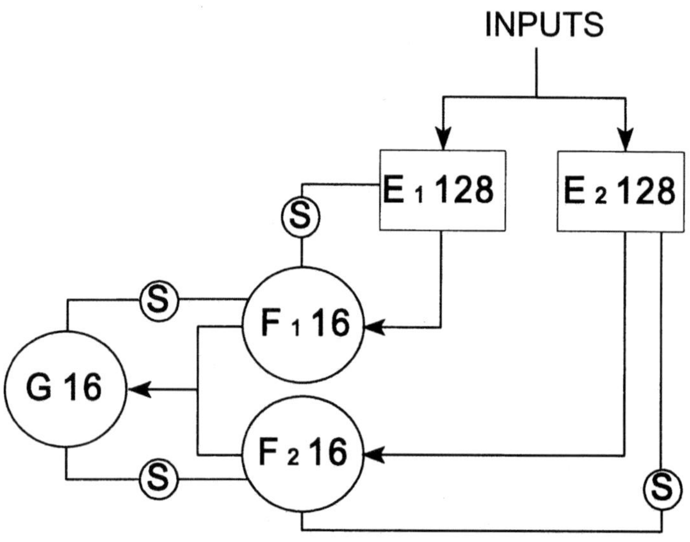

Figure 3-15: Components E, F, and G Synchronized

With isolation, multiples, and synchronization the paths through the entire complex process from A through G can be represented as shown in Figure 3-16. (Since $P_{EF} = 256$ paths, and G has 16 stations, $P_{EFG} = 256$.)

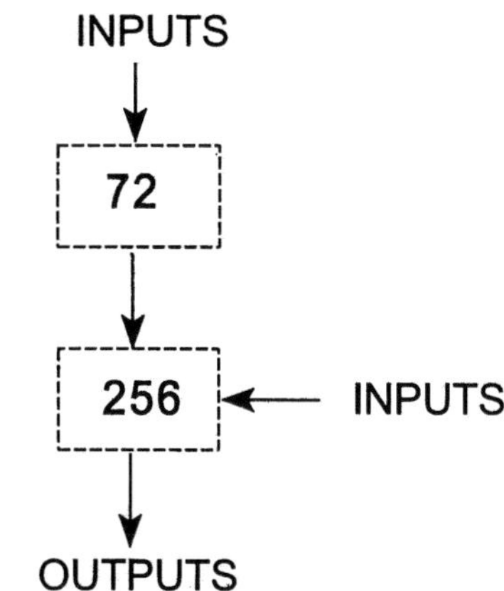

Figure 3-16: Complex Process Example, A → G

The number of process paths calculated for the entire process with isolation, multiples, and synchronization is 18,432. This is shown in the following equation.

$$P_{A \to G} = 72 \times 256$$
$$P_{A \to G} = 18,432$$

(3-7)

The total number of process paths, and therefore the potential for variation, has been reduced from 6,193,152,000 to 18,432. At the design specified output rate of 15,000 units per hour, the optimized process can run no longer than approximately 1 hour and 14 minutes before paths start repeating. This is an incredible reduction from the 82.5 years calculated at the end of Chapter 2 for the original process design.

References

1. Bajeria, H. J., Copp, R.P. *Statistical Problem Solving,* Multiface Publishing, Garden City, 1991.

2. Clements, R. B. *Handbook of Statistical Methods in Manufacturing,* Prentice-Hall, Englewood Cliffs, 1991.

3. Juran, J. M. (ed.), Gryna, F. M. (assoc. ed.) *Juran's Quality Control Handbook,* 4th edition, McGraw-Hill, New York, 1988.

Part 2

Basic Improvement Techniques

4

Problem Identification

With all of the variation that a process can generate, one might think we should never have difficulties "identifying" problems. The truth is, there are opportunities for improvement in any process, and most of us have any number of projects we could work on. Since they can't all be addressed simultaneously, one of our primary concerns must be identifying which one to work on first. This is an important step in process optimization. We must prioritize our efforts in a manner which will improve both our profitability and our probability of success. This chapter presents a few simple methods to identify which problems should be addressed first.

Pareto Charts

Dr. Joseph M. Juran introduced the concept of Pareto analysis in the 1940's. Basically, it is accomplished by organizing data into specific categories and then arranging the categories in descending order on the basis of frequency. For impact and clarity, the data is often presented in the form of a bar graph. For example, consider the data in Table 4-1:

Table 4.1: Defect Summary Sheet

Defect	Machine	Product	Quantity
scratches	101	203849	264
	102	203849	286
	103	250157	68
width under	103	203840	123
length under	101	250157	225
	103	250189	210
color	102	250189	55
thickness	103	203849	97

The defect summary sheet provides a very precise accounting of defective pieces for a manufacturing process. However, a better tool for summarizing and presenting the data is the Pareto Chart or Pareto Ranking. If we add all of the quantities contained in each defect category and plot them in descending order, we obtain Figure 4-1.

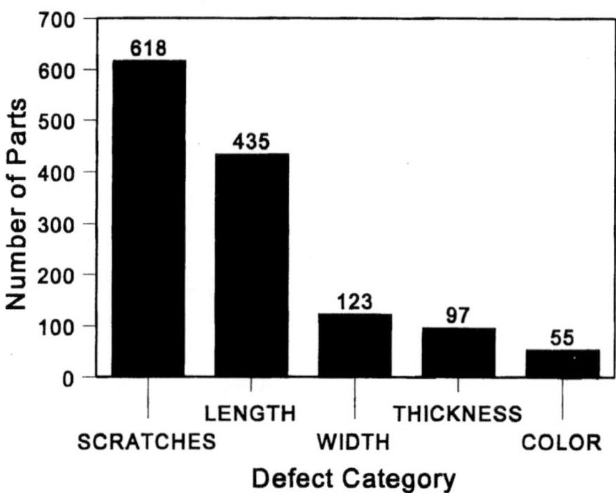

Figure 4-1: Pareto Chart of Defects

It is immediately obvious from the figure that the defect category *scratches* is the largest in this manufacturing process. Armed with this information, we could set out to improve the process by working on their elimination. While this would obviously be beneficial, it may not be the most appropriate course of action.

Cost Ranking

If we consider the defect categories closely, we can surmise that it might be possible to rework the scratched parts. Though *scratched* is a valid defect category, it may not necessarily be scrap. *Length under*, on the other hand, may not be possible to rework. All 435 pieces shown in the Pareto Chart may be scrap. Even though *length under* accounts for fewer pieces than *scratches* it may be a much costlier defect. If we take cost into consideration, we need to modify our Defect Summary Sheet. The new sheet might appear as shown in Table 4-2:

Table 4-2: Defect Summary Sheet (Cost in Dollars)

Defect	Machine	Product	Quantity	Rework Cost	Number Reworked	Scrap Cost	Number Scrapped	Total Cost
scratches	101	203849	264	1.25	250	15.00	14	522.50
	102	203849	286	1.25	268	15.00	18	605.00
	103	250157	68	1.55	68	9.00	0	105.40
width under	103	203840	123	-	0	12.50	123	1537.50
length under	101	250157	225	-	0	9.00	225	2025.00
	103	250189	210	-	0	10.25	210	2152.50
color	102	2501898	55	-	0	10.25	55	563.75
thickness	103	203849	97	-	0	15.00	97	1455.00

Although the summary sheet has become more complicated, the information depicted is more valuable. The total costs tabulated in the summary sheet are Pareto ranked in Figure 4-2.

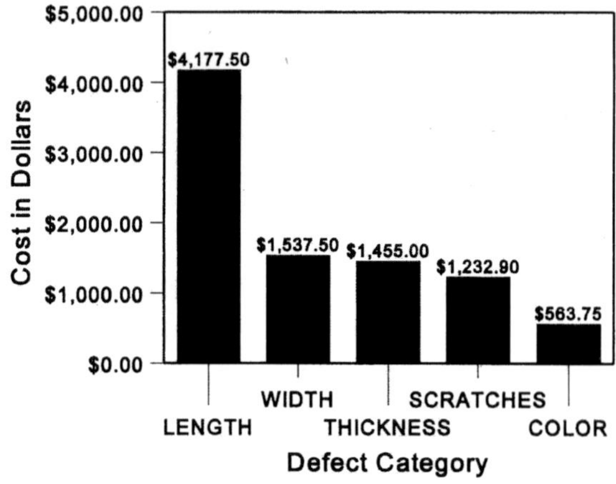

Figure 4-2: Pareto Chart of Cost

This Pareto Chart clearly indicates the costliest defect is *length under*, while the defect category *scratches* has dropped to fourth place. Obviously, eliminating *length under* defects will have the largest impact on profitability and should probably be the primary focus of improvement efforts.

Tracking and reporting the dollar volume lost will provide financial justification for process improvement expenditures, for both equipment and man-hours. This is often a prerequisite to gain management support for problem-solving. Cost determination will overcome misconceptions about which defect category is the most damaging. For example, even though *width under* is the second costliest category, in the absence of the cost chart it would probably not be perceived as a large problem. Only 123 pieces were scrapped for *width under*. Without cost data this amount would probably seem trivial when compared to the 618 pieces scratched. Additionally, the 123 *width under* pieces were scrapped, which probably means they were not visible. The 618 scratched pieces were reworked, which means they were visible. Without cost data, supervision would almost certainly select *scratches* as a larger problem since the volume of parts is larger, human resources are required to rework the condition, and the defective parts are constantly in view.

Modern quality control functions in manufacturing facilities have quality cost systems already in place. It is not unusual for comprehensive systems to tabulate hundreds of defect codes, which cost up to hundreds of thousands of dollars every month. Typically, the data is Pareto ranked by computer algorithms or purchased software. If the tracking system encompasses all of manufacturing, the quality cost report is one of the primary places to look when selecting projects to improve profitability.

Customer Requirements

Even though cost is generally the best measure to select problems for improvement or elimination, there are also other considerations in a typical manufacturing environment. For instance, the defect category *scratches* in our previous example is ranked fourth in terms of cost, but it may be the most frequent customer complaint. On this basis it is certainly a valid improvement project. We should never forget that we remain in business only as long as the customer is satisfied with our product, delivery, and price.

Suppose the defect *scratches* were selected for resolution. The table listing the cost of *scratches* as a function of manufacturing line and product is worth further investigation. It illustrates an important point in the problem solving process. If we prepare a Pareto ranking of cost due to scratches, we obtain the results illustrated in Figure 4-3.

In attempting to solve the *scratches* problem, the most natural tendency would be to select line number 102, or 102 and 101, and try to minimize the scratches created. This would be the obvious approach since the failure costs are higher on these lines. It might be more beneficial, however, to examine line 103 to determine why the costs associated with this line are so much lower than the other two. (This assumes of course, that the production volume across all of the lines is approximately the same.) The data indicates that fewer parts are scratched on this line. This comparison is known as benchmarking. If we can isolate characteristics of line 103 that tend to eliminate the formation of scratches—whether of machine, maintenance, or procedural origin—we can attempt to apply them to the other two lines.

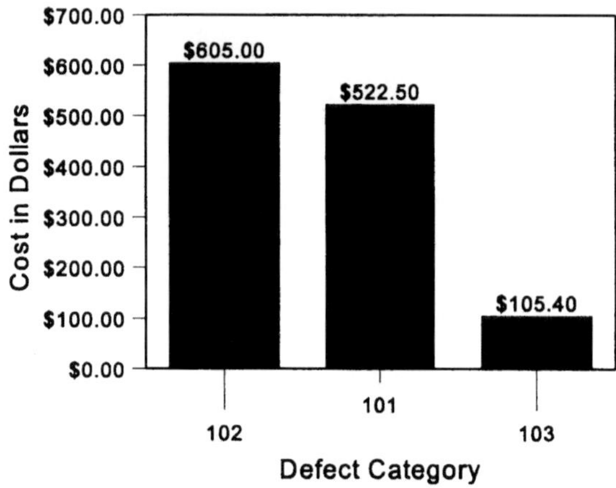

Figure 4-3: Cost of Scratches

With reference to Figure 4-3, it must also be pointed out that line number 103 has produced its scratches on a different part number than the other two lines. The product on this line, part number 250157, may account for the lower failure costs. That is, line 103 may be just as prone to scratch as the other two lines. The design of part number 250157 may make the product itself less likely to scratch. If such design features exist, they can be isolated and perhaps incorporated into the other part.

Production Inhibitors

One of the most often overlooked opportunities for improvement is a problem which tends to inhibit or restrict production output. One of the prime reasons for this is the inherent difficulty in classifying some of these problems. For instance, consider the process flow diagram shown in Figure 4-4. If the percentages shown represent the efficiencies of each operation it is obvious that operation B is near its limit of 100%. Operation B is probably restricting the flow of product. Operation A must be intentionally run at a reduced rate to prevent the production of excess in-process inventory. Operation C is running all of the product supplied by operation B, however, C has the capability to run more than twice its current volume.

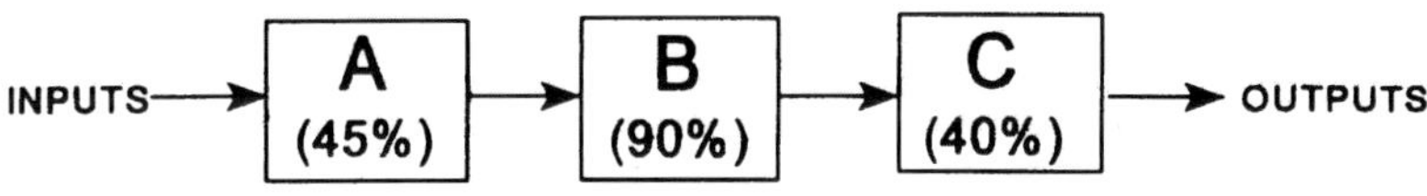

Figure 4-4: Production Inhibitor

The traditional solution to such a problem is to simply duplicate operation B. This will double productivity. All too often, however, we fail to question why an operation is restricting our process flow rates. Generally, this failure to question and examine occurs when individuals do not understand and recognize the difference between the definitions of efficiency and utilization rates. The efficiency rate is the operation output divided by its calculated capacity. This is shown in Equation 4-1.

$$Efficiency\ Rate\ =\ \frac{Operation\ Output}{Calculated\ Capacity} \tag{4-1}$$

The key phrase in the above definition is "calculated capacity." This means that allowances are made for events such as production scheduling, quality factors, machine maintenance, setup times, and operator breaks. The utilization rate of a machine or operation is defined as the ratio of operation output divided by the ultimate capacity or maximum attainable output of the machine or operation. This relationship is shown in Equation 4-2.

$$Utilization\ Rate\ =\ \frac{Operation\ Output}{Ultimate\ Capacity} \qquad (4\text{-}2)$$

The efficiency rate, since it relies on a calculated capacity, is often padded to allow for the many possible causes of downtime. When this happens, we regularly report high efficiencies and can easily convince ourselves that we are running a superb operation. The utilization rate, on the other hand, is far less deceptive. Reporting the utilization rate allows no opportunity for padding the numbers. It is not uncommon for many manufacturing facilities to report very high efficiencies, sometimes at or near 100%, when actual machine utilization rates are 35%-45%.

Operations such as these are prime opportunities for improvement. Increasing the utilization rate of operations will prevent the needless:

1) purchase of capital equipment
2) expansion of the physical plant
3) hiring of additional operators
4) hiring of additional support personnel
5) duplication of tool inventory
6) expansion of repair part/maintenance inventory

In short, increasing machine utilization rates will improve productivity and improve quality while simultaneously lowering costs.

Returning to Figure 4-4, instead of duplicating operation B, we should first question if the efficiency and the utilization rates of the machine are similar. If the efficiency is high and the utilization is low we should examine the operation in detail. What is the cause of the low utilization rate? Is it the scrap rate of this operation? Is it excess downtime caused by material shortages? Is it maintenance or reliability related? Is it caused by unduly long setup times? There are many obvious possibilities. Even though the machine is running at 90% efficiency rate, it may quite possibly be grossly underutilized. By addressing the causes of low utilization rates in an operation, we can frequently gain enormous improvements in our process without investing in additional equipment, facilities, and personnel.

Personnel Considerations

Another critical factor to evaluate when identifying and selecting a problem for resolution is the probability of actually solving the problem. While this is often difficult to assess, it is important to select problems commensurate with the talents of the people who will be working toward resolution of the problem. Few things are more demoralizing than giving a group a problem that is beyond its ability to solve. When a problem is selected, it is important to assign personnel who have the necessary training and/or background such that they have a reasonable chance of actually solving the problem.

References

4. Berger, R. W. (ed.), Pyzdek, T. (ed.) *Quality Engineering Handbook,* ASQC Quality Press, Milwaukee, 1992.

2. Juran, J. M. (ed.), Gryna, F. M. (assoc. ed.) *Juran's Quality Control Handbook,* 4th edition, McGraw-Hill, New York, 1988.

3. Stevenson, W. J. *Production /Operations Management*, 4th edition, Richard D. Irwin, Boston, 1993.

5

Basic Problem Solving

This chapter will examine a few basic problem solving techniques. They sound easy and they are. It's unfortunate, but these techniques are frequently either not utilized or underutilized. They are basic and should not be relegated only to a quality engineer or quality manager. They belong in everyone's "toolbox" if process optimization is a company goal. Many works are available which cover the techniques we are going to discuss in this chapter. Readers are encouraged to review the references listed at the end of the chapter.

Ishikawa Diagram

This diagram, also known as a Cause and Effect or C&E Diagram, is one of the easiest and most useful tools for problem solving. Developed by Dr. Kaoru Ishikawa of the University of Tokyo in 1943, it is especially suited to the "brainstorming" process, where a team of personnel work together to identify possible causes of problems. A single problem, or a group of problems, is classified as the outcome or effect(s) of many potential causes. This relationship is illustrated Figure 5-1.

Figure 5-1: Cause and Effect

Rather than randomly listing all of the possible causes of a problem, the C&E Diagram divides them into categories. For instance, a C&E Diagram usually includes additional lines or branches for categories such as Man, Methods, Materials, and Machine. This is shown in Figure 5-2.

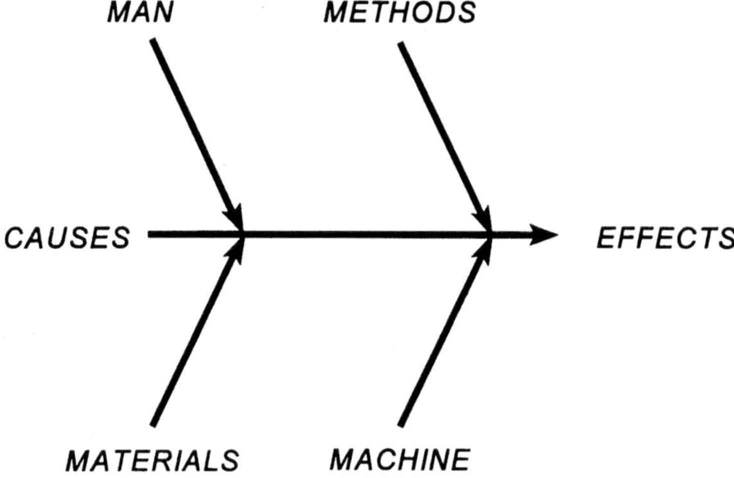

Figure 5-2: Cause and Effect Diagram

The diagram is not limited to the categories shown. It is not unusual to find diagrams with additional categories such as Environment or Time. Any number can be added, as applicable to the problem at hand. Because of its appearance with category lines added, the Ishikawa Diagram is often referred to as a "fishbone diagram."

The example in Figure 5-3 was the result of a problem solving team formed to eliminate warping of the paper/composite lining found on the inside of caps used on food containers. The tendency of a cap liner to warp, while not important after screwed on a container, is a very important characteristic prior to installation. If the warp, or curl, was severe enough the diameter of the liner would effectively be reduced and it would fall out of the cap prior to reaching the container.

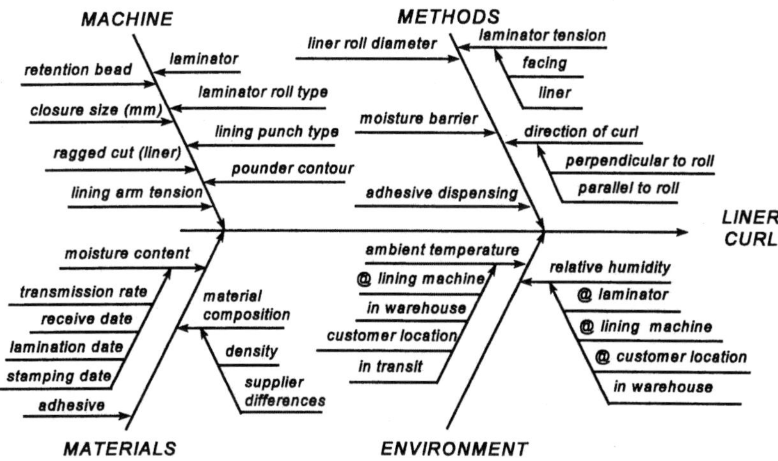

Figure 5-3: Cause and Effect Diagram

The C&E Diagram shown above illustrates the interrelationships between the potential causes and the specific problem selected. Without the benefit of a group brainstorming this process, there would not have been quite so many potential causes identified. This diagram was used as the basis for selecting factors for inclusion in a designed experiment, which ultimately led to the elimination of the problem and associated customer returns valued at almost $2,000,000 per year.

The Cause and Effect Diagram is simply one of the most important and useful steps in the problem solving/process optimization process. It is often overlooked because individuals tend to think they know the cause of the problem being investigated. While there is a great amount of flexibility in precisely how the diagram is completed in terms of categories and causes, there are a few "ground rules" which should be followed in order to obtain useful diagrams. These are:

1) The C&E Diagram should be completed by a group.
2) All causes volunteered are valid and must be recorded.
3) Recording should continue until all members have no ideas.
4) All causes should be discussed after completion of the diagram.
5) Group members should vote to determine most likely causes.

Concentration Diagrams

Another very simple, and often overlooked problem solving tool is the concentration diagram. This is nothing more than a simple recording of defect location on a sketch of the product. If a specific defect tends to occur frequently in specific locations, this can provide enough information to isolate and correct the primary, or root cause of the defect.

Example 5.1

A manufacturer of headlamps was experiencing a low level of customer returns for hermeticity failures. With this type of defect, the lens of the headlamp was not forming a watertight seal to the reflector assembly. As returns were received, the Quality Engineer assigned to determine root cause recorded the failure locations on a sketch of the headlamp configuration. Each assembly was carefully dissected and examined after initial data for leak location and pressure differential at failure was recorded. Brainstorming the problem had led to the selection of the "middle of the long side" as the most likely failure location (Figure 5-4). This location was predicted because of the construction of the assembly. It was molded from plastic, and the long sides were inherently the most difficult to maintain dimensionally.

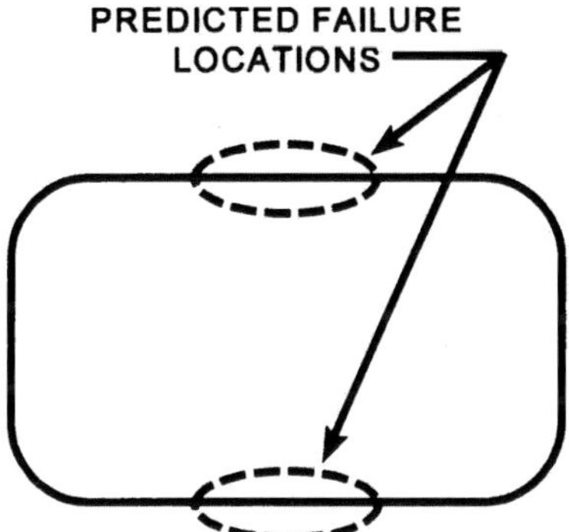

Figure 5-4: Front View of Headlamp Assembly

As the molded reflector portion of the headlamp assembly was ejected from the molding press and allowed to cool to ambient temperature, the two long sides had a tendency to distort more than the corners of the assembly. Since the adhesive which bonded the lens to the reflector was robotically dispensed according to a computer program, the dimensions of the reflector were critical. It was reasonable to assume, then, that if a failure were going to occur, distortion of the reflector would cause failures due to the improper placement of adhesive.

It took only a few customer returns for a failure pattern to emerge. Surprisingly, it did not look like expected. Most hermeticity failures were concentrated in one corner of the assembly, as shown in Figure 5-5.

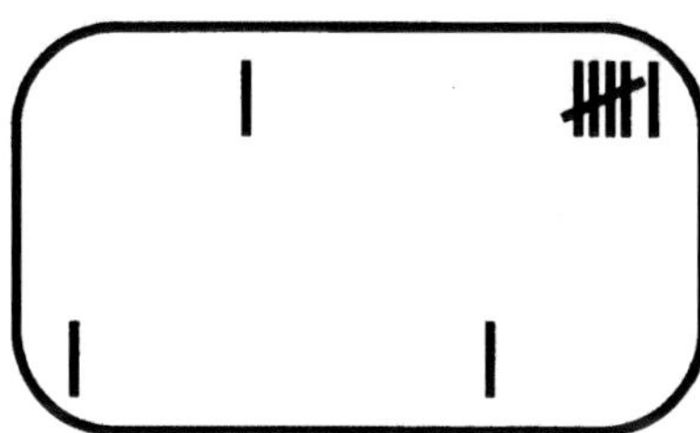

Figure 5-5: Concentration Diagram of Failures

With this information, the Quality Engineer observed the assembly operation closely. The operators had to stretch across the table used for robotic dispensing of the adhesive. They did this as they were assembling the lens onto the reflector. As a consequence, they tended to "plow" or push the adhesive in the reflector assembly out of position. This was the root cause of the hermeticity failures in the assembly.

The corrective action for this failure mode was to remove the operator from the assembly process. An automatic assembly machine was constructed to precisely place the lens onto the reflector assembly. Instead of assembling the components, the operator's task was to load the lenses into the positioning device, and remove the assembled components for further processing.

Example 5.2

A metal cap manufacturer was experiencing problems with one line of product. The process consisted of stamping metal shells, followed by a few intermediate operations and finished with an operation that rolled the cut edge of the shell over toward the inside of the cap. An intermittent defect was appearing on the edge of the cap, specifically, a portion of the shell would crush in the final stage of the manufacturing process.

All of the caps were decorated with customer's artwork, that is, lithography was in place on the shell prior to the forming operations. In spite of this, the failure appeared to demonstrate no discernable pattern. Shells were stamped from 3 punches on one press concurrently. Punch numbers were included on the lithography and all three caps were failing, although the frequency rate for punch number one was less than the other two punches. An illustration of the failure mode is shown in Figure 5-6.

After many hours of investigation of machine setup, processing parameters and metallurgical properties, no apparent remedy was in sight. It was possible to alter process parameters to minimize, but not eliminate the defect from production.

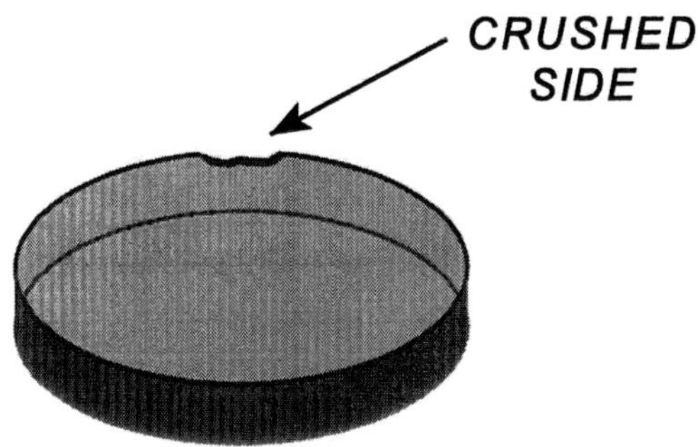

Figure 5-6: A Stamping Defect

As failures continued to accumulate sporadically, they were collected for analysis. A decorated sheet was obtained prior to the stamping process and the progression of the sheet through the punch press was examined. The layout of the sheet is shown in Figure 5-7.

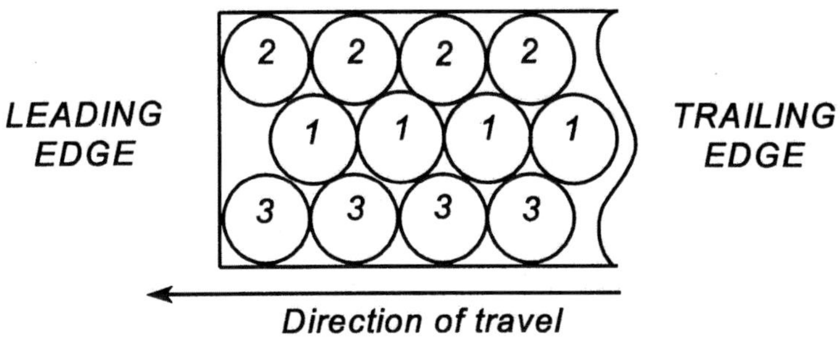

Figure 5-7: Sheet Layout

The spacing of the three punches on the press determined the stamping pattern. As the sheet was fed into the press, the first shell stamped was the number 1 shell closest to the leading edge. In the next progression all three punches stamped shells. This pattern continued until the end of the sheet which contained 48 shells. The progression of the sheet is partially illustrated in Figure 5-8.

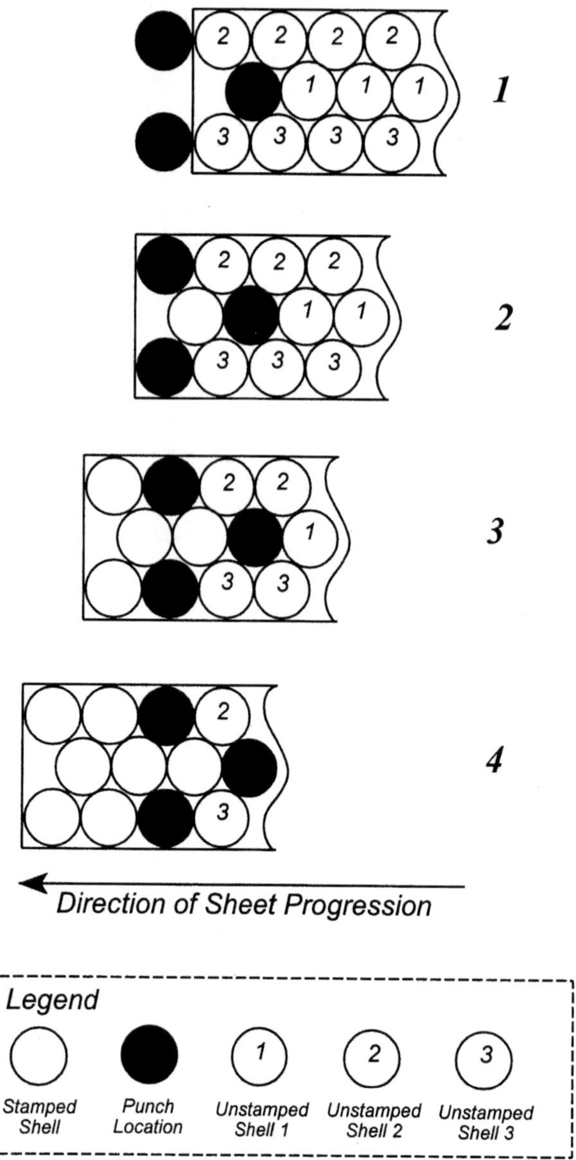

Figure 5-8: Sheet Progression Through Stamping Punches
(First Four Progressions Shown)

A concentration diagram was constructed on a decorated sheet by orienting the failed caps with the unstamped caps on the decorated sheet. This was a simple matter accomplished by aligning the lithography. In order to obtain a diagram which represented the bulk of the caps stamped, a progression location was chosen which coincided with the third progression (Figure 5-8). This did not represent either the first shell stamped at the leading edge, or the last two shells stamped at the trailing edge. The concentration diagram obtained is shown in Figure 5-9.

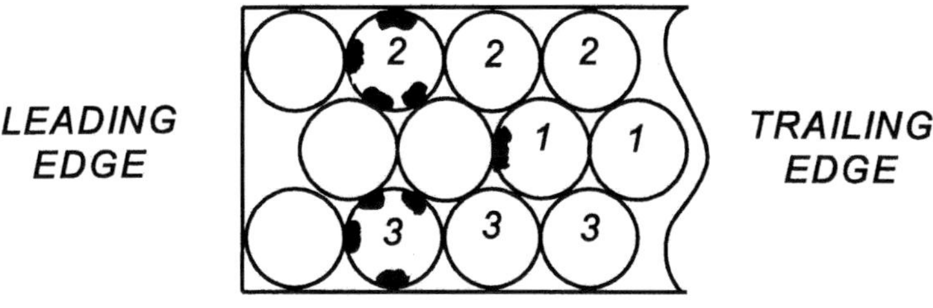

Figure 5-9: Concentration Diagram

After plotting a failure locations on the diagram for a few days, distinct concentrations began to appear for each shell. These are represented by the darkened areas in Figure 5-9. For example, shell number one failures were almost all located toward the side of the sheet marked "leading edge." Undoubtedly, this was because the shell to the left of number one had been punched out in the previous progression. The material remaining at this location, that is, the "web" of the sheet was too thin. There was not a sufficient amount of steel remaining between the layout of the two shells for the punch to produce a clean "cut." The "web" for this interface is illustrated in Figure 5-10.

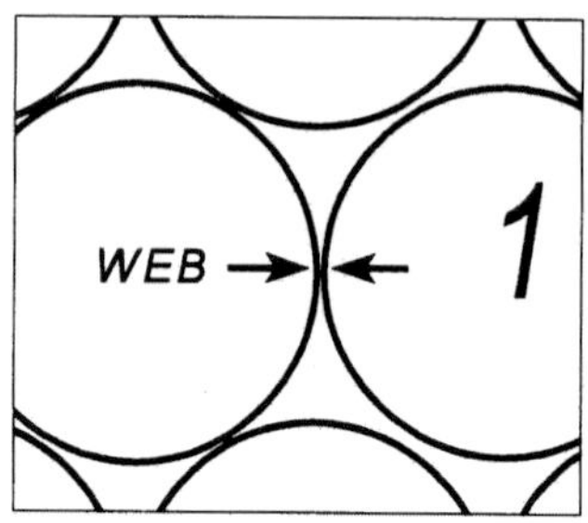

Figure 5-10: Web

The majority of failures for shells number two and three were likewise located either at the edge of the sheet, or adjacent to locations that were stamped out in prior progressions.

Because the web was too thin in all these locations, it lacked the structural rigidity required to remain in place as the punch was stamping out the next shell. Consequently, a portion of the web was pulled into the next shell being stamped, increasing the height of the shell in localized areas. This high portion of the shell, illustrated in Figure 5-11, was out of specification for the operation that rolled the bead on the end product. There was too much material in these areas and, unless the setup was nearly perfect, the bead operation crushed the shells at these locations.

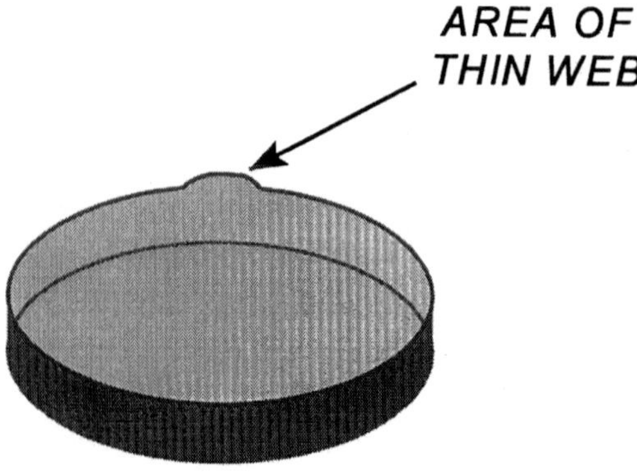

Figure 5-11: Thin Web Excess Material

Run Charts

Another basic tool is the run chart. This is simply a line chart of the process or characteristic under investigation. By plotting the data, and annotating the chart with significant events, we can determine over time if our efforts are producing the desired results. A run chart requires no special training and will indicate if the process is:

        1) getting better
        2) remaining the same
        3) getting worse

This tool also has the distinct advantage of being universally understood. For this reason, it is a excellent method to communicate performance plant wide.

Example 5.3

Consider the chart shown in Figure 5-12, taken from a manufacturing process which applied a plastic-like material on a product.

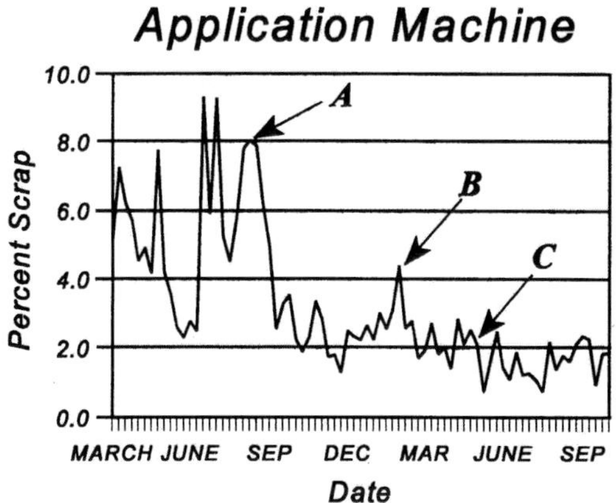

Figure 5-12: Run Chart

The three events, denoted by the letters A, B, and C each represent distinct attempts to improve the process. The effectiveness of each step was evaluated by determining the amount of decrease in the percentage of product scrapped. Notice that the chart represents one and one-half years of production. The improvement efforts span a one year period. These efforts were successful in reducing the average scrap rate from just under 6% to less than 2%. This scrap reduction was equivalent to a savings of approximately $250,000 per year.

The improvements consisted of procedure development, operator training, and minor expenditures on equipment to control the application temperature of the materials. The total expenditure on equipment and materials was less than $10,000.

One of the keys to the success of the run chart is visibility. All personnel tend to take interest in a process that is being charted, both as a matter of pride if they are involved with the process and because the chart demonstrates management interest. When performance criteria and goals are clearly posted, people tend to perform at higher levels.

References

1. Ishikawa, K. *Guide to Quality Control,* Asian Productivity Organization, Tokyo, 1982.

2. Juran, J. M. *Juran on Quality Improvement,* Juran Enterprises, New York, 1981.

6

SPC Review

This chapter provides a review of statistical process control principles. The review is preparation for Chapter 7 where we examine the application of SPC to complex processes. It is intended only as a review of SPC basics. Even though these techniques have been with us for almost 70 years, many individuals still do not understand the principle that SPC is founded on. Some think the output from their manufacturing process must be "normal" for SPC to work. As we shall see, this signifies a fundamental misunderstanding of SPC theory.

Statistical process control techniques are the best way to monitor a manufacturing process. They will indicate when the process is unstable, that is, when there is assignable cause variation affecting the process. Finding the cause of the variation, however, is seldom easy. Often, locating sources of variation requires implementation of the problem solving techniques discussed earlier.

In preparing this review, it is not my intention to cover statistical process control techniques in any real degree of detail. My intent is merely to emphasize the fundamentals and explain why statistical process control works. There are a variety of excellent works on the subject matter already, including *Economic Control of Quality of Manufacturing*

Processes by Walter A. Shewhart and *Statistical Process Control* by Eugene L. Grant and Richard S. Leavenworth. You should consult these and other references for a more detailed discussion if you are applying statistical process controls to your manufacturing process.

Types of Distributions

Manufacturing processes may produce outputs distributed in a variety of shapes. There are, however, just a few distribution patterns which represent the majority of manufacturing processes. These distributions are illustrated in Figure 6-1.

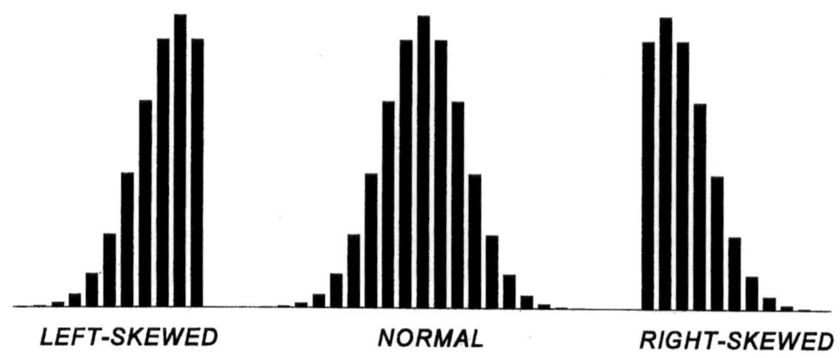

Figure 6-1: Typical Manufacturing Distributions

The left-skewed and the right-skewed distribution can occur when a process output is constrained by some sort of barrier, either physical or computational. For example, measure of a process yield might result in a left-skewed distribution. While process yield could approach 100%, it could not go beyond. A right-skewed process might occur if we are attempting to minimize a non-negative characteristic such as push-in force for a connector.

The vast majority of manufacturing processes produce outputs which are distributed in a normal fashion. This distribution is characterized by symmetry and a bell-shape For reasons we shall examine, a basic understanding of normal distribution properties is important for the successful application of statistical process control.

The Normal Distribution

Since most manufacturing processes produce outputs distributed in a normal fashion, it is useful to examine the properties of this type of distribution. In graphical representations of data, the width of the bars, or cells, in a distribution is determined by the number of data points available. Considering the normal distribution in depicted in Figure 6-1 once again, the same distribution would appear as illustrated in Figure 6-2 if additional data points were included in the distribution plot.

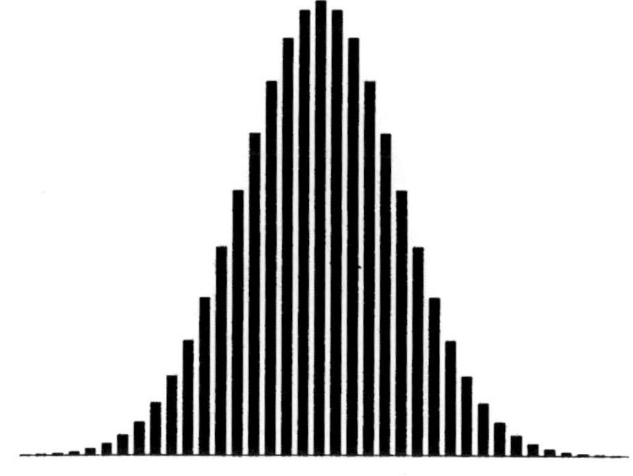

Figure 6-2: The Normal Distribution

As more data is included, the distribution gains more cells and the distinction of each bar becomes more difficult to see. It then becomes convenient to represent the distribution as a smooth curve rather than as a bar graph. The curve representing the normal distribution shown above is illustrated in Figure 6-3.

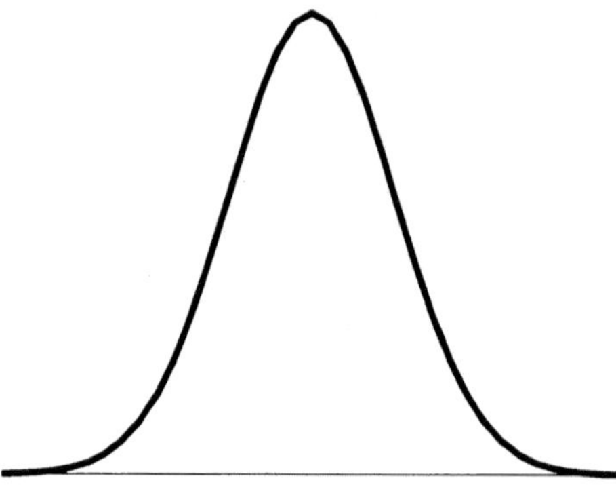

Figure 6-3: The Normal Curve

The normal curve represents the distribution plot with an infinite number of data points. The mathematical equation for the normal curve is shown below:

$$y = \frac{1}{\sigma \sqrt{2\pi}} \, e^{-(X-\mu)^2 / 2\sigma^2}$$

$$\begin{aligned}
where: \; e &= 2.718 \\
\pi &= 3.141 \\
x &= individual\ measured\ value \\
y &= height\ of\ response \\
\mu &= population\ average \\
\sigma &= population\ standard\ deviation
\end{aligned}$$

(6-1)

Two parameters are needed to quantify the normal distribution. These parameters, which are evident in Equation 6-1, are the central tendency or population average, and the process dispersion or standard deviation. Once these values are inserted in the equation, the height of the normal curve, y, is defined for all values of x. The population average is found by summing all of the measured values and dividing the total by the number of values summed. This is shown in Equation 6-2.

$$\mu = \frac{\Sigma\, x_i}{n}$$

where: μ = *population average*   (6-2)

Σ = *mathematical symbol denoting sum*

x_i = *individual measured value*

n = *number of values*

The standard deviation is a bit more complicated to calculate, however, it is a critical parameter of a normal distribution. The formula for the standard deviation of a population is illustrated in Equation 6-3. Most modern calculators will compute this parameter with minimal effort.

$$\sigma = \sqrt{\frac{\Sigma\,(x_i - \mu)^2}{n}}$$

where: σ = *population standard deviation*

Σ = *mathematical symbol denoting sum*   (6-3)

x_i = *individual measured value*

μ = *population average*

n = *number of values*

Changing the population average, μ, will change the position of the normal curve on the x axis, but will not affect its shape. This is illustrated in Figure 6-4 for increasing values of μ. The significance of this relationship is important not only for the understanding of statistical process control, but also in process optimization efforts involving designed experimentation. As we shall see, SPC techniques will "signal" when a process average has shifted. In reaction to this shift, we will try to identify the cause and remove it from our process. In designed experimentation, we try to identify the factors which can cause shifts and use them to target process output exactly where we want it to be.

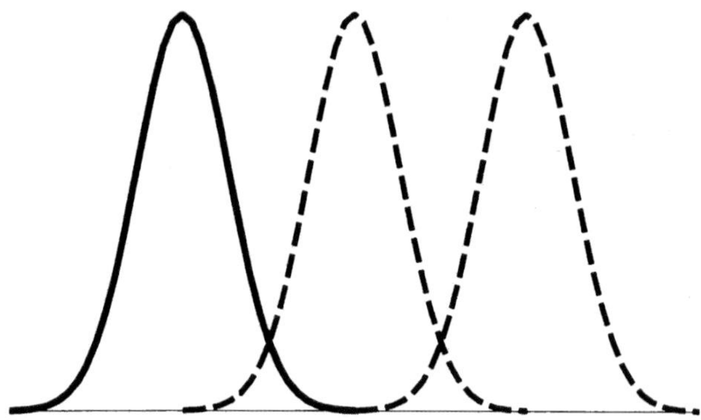

Figure 6-4: Changes in Population Average

Changing the standard deviation, σ, will change the shape of the normal curve, but will not affect its position. This is illustrated in Figure 6-5 for increasing values of σ.

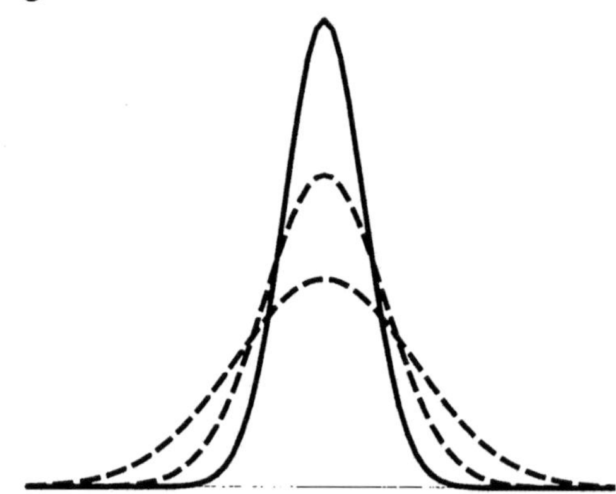

Figure 6-5: Changes in Standard Deviation

As was the case with the process average, statistical process controls will tell us when one or more factors have acted upon our process and caused the standard deviation to change. Likewise, when we examine designed experimentation we shall learn how to identify the factors which affect standard deviation and use them to improve our process by reducing variation.

All of the curves depicted in Figure 6-5 are normal. The changes in shape are strictly due to the changes in standard deviation. This illustrates an important property of the normal curve. Even though the shape may change, the total area underneath the curve is always equal to 100% or one (1). The relationship between μ, σ, and the area under a normal curve is illustrated in Figure 6-6.

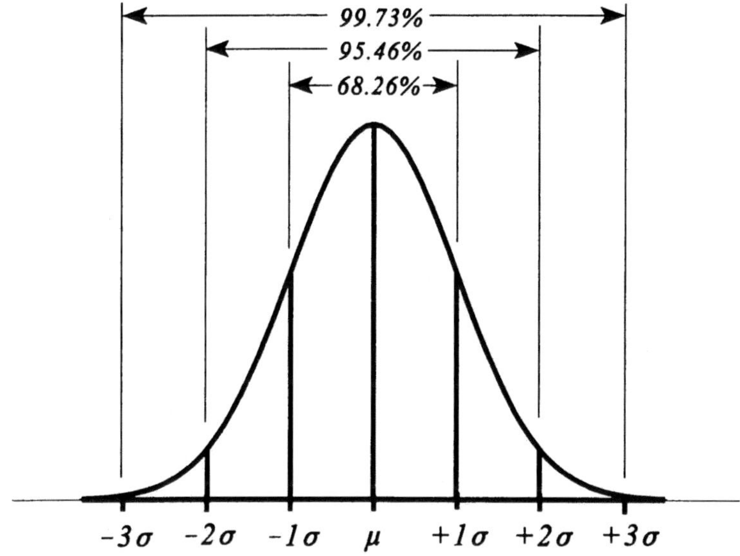

Figure 6-6: Area Properties of the Normal Curve

From the illustration, we can see that 68.26% of the distribution values will lie within $\pm 1\sigma$ of the mean, 95.46% will lie within $\pm 2\sigma$ of the mean, and 99.73% will lie within $\pm 3\sigma$ of the mean. These areas are constant for all normal curves. One additional point, the normal curve is asymptotic; it continuously approaches, but never actually reaches the x-axis. It is possible to calculate areas under the curve for additional standard deviations away from the mean such as 4σ, 5σ, etc. We shall examine the principle behind statistical process control in the next section. It will become evident why the properties of the normal curve are so important.

The Central Limit Theorem

The central limit theorem, in simple terms, states:

> *If samples are drawn from a process (the sample size must be greater than or equal to four[1]), and the average of each sample is plotted, the distribution of averages will be normal, irrespective of the distribution of the individuals.*

This is an extremely useful theorem, for it indicates we will always know the shape of the distribution of averages. Since we always know the shape, we may develop schemes for charting and reacting to the plotted points. In other words, even though some manufacturing processes do not produce normal distributions, we will always obtain a normal distribution if we draw samples of the process output, and plot the average values of the samples drawn. The Central Limit Theorem, or CLT, is illustrated in Figure 6-7 for a right-skewed process.

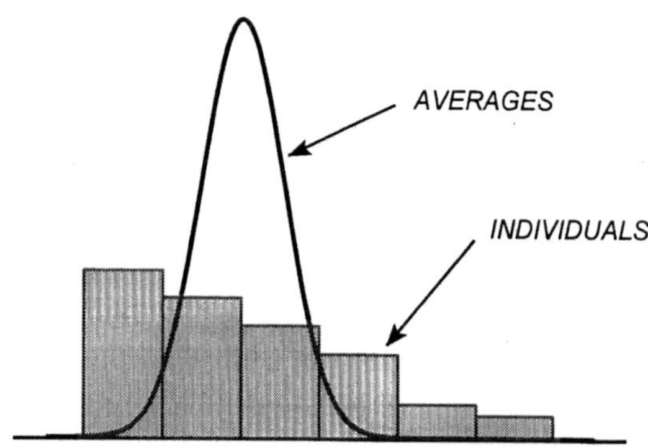

Figure 6-7: CLT for Right-Skewed Distribution

---

[1]Some references indicate the minimum sample sizes should be as large as ten.

The Central Limit Theorem is illustrated in Figure 6-8 for a process with a normal distribution of individuals.

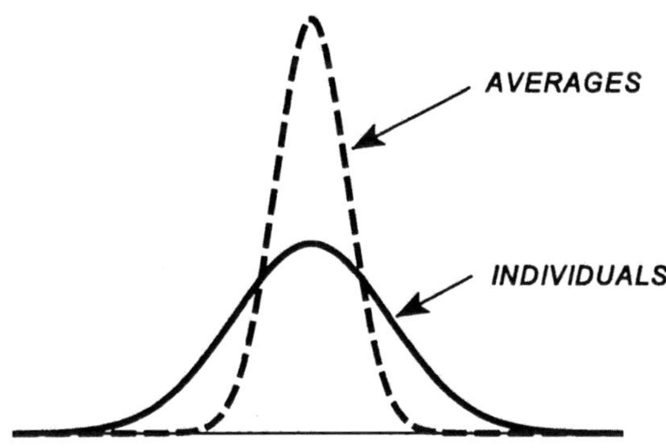

Figure 6-8: The Central Limit Theorem

Once again, the importance of the Central Limit Theorem is the prediction of a normal distribution of averages, irrespective of the shape of the distribution of individuals. Since we know the properties of the normal distribution in terms of central tendency and dispersion, we may make predictions about the distribution of plotted averages in terms of the average of the averages (central tendency) and the standard deviation of the averages (dispersion). This is the basis of the $\bar{x} - R$ chart so common in manufacturing environments.

In Figure 6-8 we saw a distribution of averages plotted on the same scale as a distribution of individuals. Since 99.73% of the individual values are expected to lie within the range of $\mu \pm 3\sigma$, it follows that 99.73% of the averages should lie within the range of $\bar{\bar{x}} \pm 3\sigma_{\bar{x}}$, where $\bar{\bar{x}}$ represents the average of the plotted averages and $\pm 3\sigma_{\bar{x}}$ represents the six standard deviation spread of the plotted averages. Upper and lower control limits for the $\bar{x} - R$ chart are established at $\bar{\bar{x}} \pm 3\sigma_{\bar{x}}$, as shown in Figure 6-9.

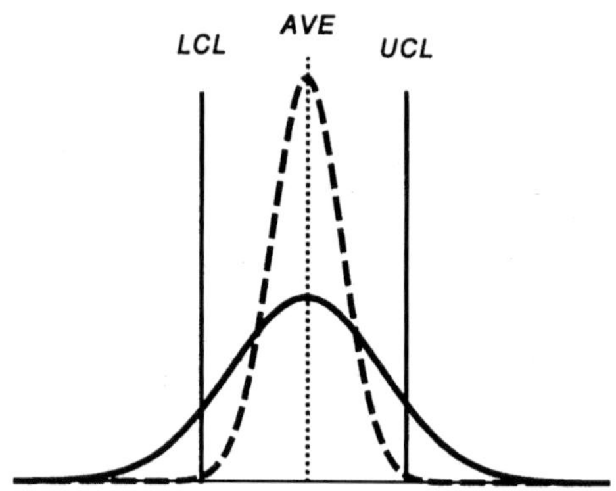

Figure 6-9: Upper and Lower Control Limits

The practical value of the control limits lies in the realization that 99.73% of the averages plotted will be between the upper and lower control limits when the samples are drawn from a stable population. There is only a .0027 probability (1-.9973) that averages would be plotted above or below the control limits. In other words, only 27 plotted points in 10,000 would naturally fall either above the upper control limit (UCL) or below the lower control limit (LCL). This is roughly 1 in 400. Plotting a point beyond the control limits indicates the process is not stable. An unstable processes is termed "out of control."

$\overline{x}$ - *R* Charts

The most widely used control chart is called the $\overline{x}$ - *R* chart. It has this name because it consists of two simultaneous plots, one for the central tendency or average of the sample drawn, $\overline{x}$ - *R*, and the other for the dispersion or range within each sample, R. To produce the $\overline{x}$ - *R* chart we begin by drawing samples from the process being evaluated. The data in each group of samples is summed and the average is calculated. The range of the sample group is also calculated by identifying the largest and smallest individual value, and determining the difference between them. The average and range are then plotted on a graph in time order. After enough data is plotted, usually 125 individual measurements, control limits are calculated from predefined formulas, as shown in the references at the end of this chapter. The control limits are drawn on both of the plots and the process performance is compared to these limits. If the process is stable, the plotted averages and ranges will fall within the control limits, and will not display any patterns or trends except those which might be expected from a normal distribution. For illustrative purposes, the a portion of a typical $\bar{x}$ - R chart containing the first five plotted points is shown in Figure 6-9.

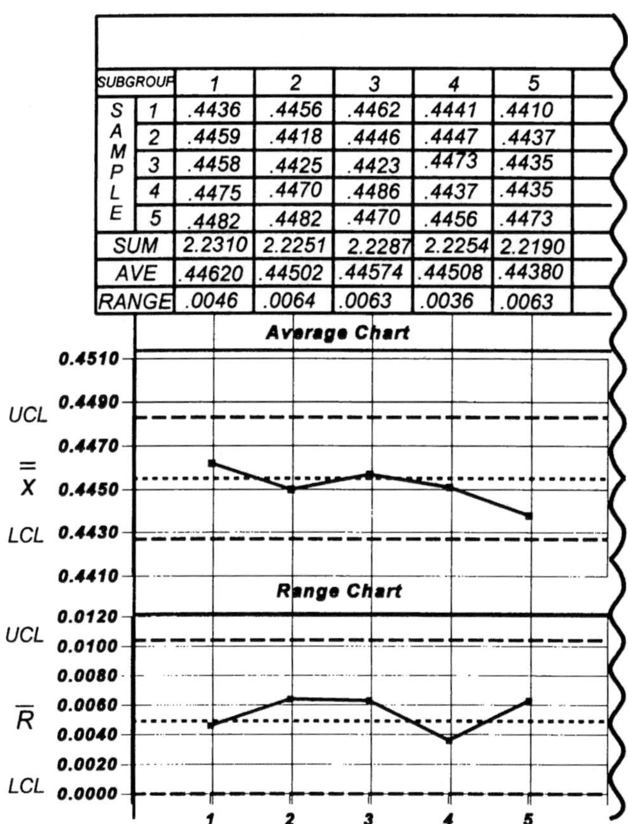

SUBGROUP		1	2	3	4	5
S	1	.4436	.4456	.4462	.4441	.4410
A	2	.4459	.4418	.4446	.4447	.4437
M	3	.4458	.4425	.4423	.4473	.4435
P	4	.4475	.4470	.4486	.4437	.4435
L E	5	.4482	.4482	.4470	.4456	.4473
SUM		2.2310	2.2251	2.2287	2.2254	2.2190
AVE		.44620	.44502	.44574	.44508	.44380
RANGE		.0046	.0064	.0063	.0036	.0063

Figure 6-9: Typical $\bar{x}$ - R Control Chart Format

I-MR Charts

Another important control chart is the Individuals and Moving Range (I-MR) Chart. This type of chart was developed and is normally used for processes where only one reading is possible. One example of such a process would be measuring the pH of a homogenous mixture in a batch operation. For this chart, individual values are plotted along with the moving range. The moving range is simply the difference between the present and previous individual values. In the next chapter we will examine I-MR applications in conjunction with $\bar{x} - R$ charts. A portion of a completed I-MR chart is shown in Figure 6-10. The equations for calculating I-MR control limits are shown in the references at the end of this chapter.

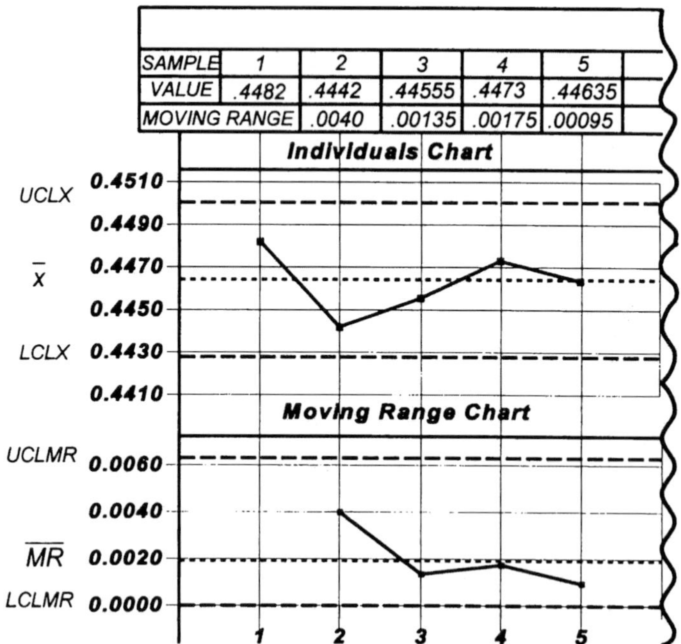

SAMPLE	1	2	3	4	5
VALUE	.4482	.4442	.44555	.4473	.44635
MOVING RANGE		.0040	.00135	.00175	.00095

Figure 6-10: Typical I-MR Control Chart Format

Types of Variation

There are two different types of variation we must distinguish when evaluating control charts. The first type, random variation, is the variation which causes deviations from average in a predictable fashion. This is the deviation which causes the normal curve to attain its characteristic bell-shape. With only this variation present in our processes, the processes will be stable over time. Their output therefore, is predictable over time. The second type is referred to as assignable cause variation. This type of variation causes our processes to act erratically. The control chart is a tool for determining the presence of assignable cause variation by documenting when the distribution of averages deviates from normal. As implied by the name, we can isolate the source of this variation and "assign" the cause of the instability in the process to the identified source. Once located, we must then remove the root cause of this type of variation to insure it no longer affects our process. Once all sources of assignable cause variation are removed the process will remain stable.

Out of Control Conditions

The $\bar{x} - R$ control chart is useful because it demonstrates the stability, or lack of stability in a process. In a stable process, we expect the distribution of averages to plot on our chart in a normal fashion. This is powerful information which can be analyzed in a variety of ways. For example, from our knowledge of the normal distribution, approximately:

1) 99.7% of the averages should plot within $\pm 3\,\sigma_{AVE}$
2) 95.5% of the averages should plot within $\pm 2\,\sigma_{AVE}$
3) 68.3% of the averages should plot within $\pm 1\,\sigma_{AVE}$
4) 50% of the average should be above $\bar{\bar{x}}$
5) 50% of the averages should be below $\bar{\bar{x}}$

When these conditions, or derivations of these conditions are not true, we know the process is no longer consistent. While others are possible, the three basic conditions on the $\bar{x}$ chart or on the range chart which identify unstable or out of control processes are:

1) one point beyond a control limit
2) seven points successively either increasing or decreasing
3) seven successive points either above or below the average

Two Measures of Process Capability

While the $\bar{x}$ - R plots averages to monitor the process' stability, and the I-MR chart plots individuals, we often need information about the distribution of individuals to know if a process is producing acceptable parts. Because this is such an integral part of process optimization, we shall briefly review the relationships between some $\bar{x}$ - R control chart parameters and the distribution of individual parts. *For all of the following relationships, we must first verify that the distribution of individuals is normal.*

The C_p Index

The C_p index is a measure of whether or not the process can meet the specification requirements when properly targeted. It is simply the specification width divided by the process dispersion as shown below:

$$C_p = \frac{USL - LSL}{6\ \hat{\sigma}_{ind}}$$

where C_p = process capability

USL = upper specification limit

LSL = lower specification limit

$\hat{\sigma}_{ind}$ = estimated standard deviation of individuals

(6-4)

The estimate of the standard deviation of individuals, $\hat{\sigma}_{ind}$, can be calculated from simple control chart parameters detailed in this chapter's references.

If $6\hat{\sigma}_{ind}$ is less than the width of the engineering specification, the C_p ratio will be larger than 1 and our process is capable of producing acceptable parts. As illustrated in the following figure, this does not mean that we are actually producing acceptable parts. It only means we are theoretically capable of producing acceptable parts. For instance, the distributions illustrated in Figure 6-11 all have the same process dispersion and the same calculated C_p value.

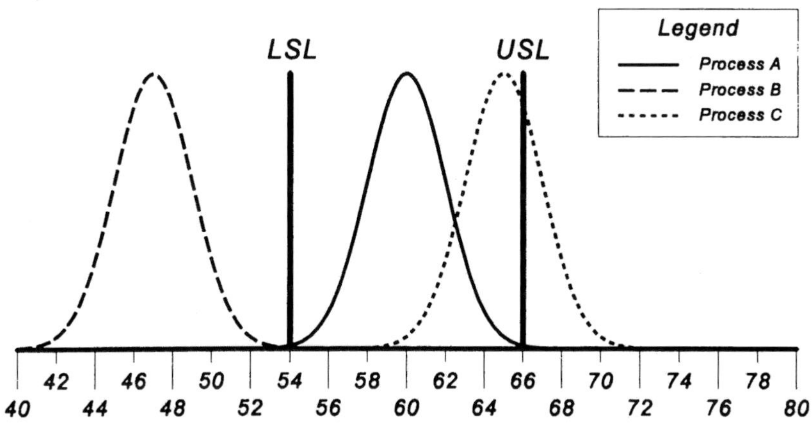

Figure 6-11: Various Distributions with Equal C_p Values

For this illustration, the specification is 60 ± 6. Each process has an estimated standard deviation of individuals equal to 2, and the C_p calculated for each process is equal to one. Each process, A, B, and C, would be capable of meeting the engineering specification if it were properly targeted at 60. (Note: A C_p value equal to 1 with a perfectly targeted process implies 99.73% of the product would be in specification.) This illustrates a shortcoming of the C_p index as a measure of process capability. It does not account for centering or targeting of the process. This problem is avoided by use of a different measure of process capability, the C_{pk} index.

The C$_{pk}$ Index

The C$_{pk}$ index is a measure of process capability which takes process centering into account. As shown in Equation 6-5, the equation for C$_{pk}$ is dependent upon the location of the process average and the nearest specification limit.

$$C_{pk} = \frac{\hat{\mu} - LSL}{3\,\hat{\sigma}_{ind}} \quad or = \frac{USL - \hat{\mu}}{3\,\hat{\sigma}_{ind}}$$

$$where\ C_{pk} = process\ capability$$
$$USL = upper\ specification\ limit$$
$$LSL = lower\ specification\ limit$$
$$\hat{\mu} = estimated\ process\ average$$
$$\hat{\sigma}_{ind} = estimated\ standard\ deviation\ of\ individuals$$

(6-5)

If the process average ($\hat{\mu}$ is estimated by $\bar{\bar{x}}$) is closer to the lower specification limit, or LSL, we simply subtract the LSL from the process average and divide the result by $3\,\hat{\sigma}_{ind}$. If the process average is closer to the upper specification limit, or USL, we subtract the process average from the USL and divide the result by $3\,\hat{\sigma}_{ind}$. The processes illustrated earlier with identical C$_p$ values are repeated in Figure 6-12. The calculation of C$_{pk}$ values for these processes is shown below.

$$For\ Process\ A: C_{pk} = \frac{\hat{\mu} - LSL}{3\,\hat{\sigma}_{ind}} = \frac{60 - 54}{3 \times 2} = 1.0$$

$$For\ Process\ B: C_{pk} = \frac{\hat{\mu} - LSL}{3\,\hat{\sigma}_{ind}} = \frac{47 - 54}{3 \times 2} = -1.17$$

$$For\ Process\ C: C_{pk} = \frac{USL - \hat{\mu}}{3\,\hat{\sigma}_{ind}} = \frac{66 - 65}{3 \times 2} = .17$$

(6-6)

The C_{pk} calculations shown require a few brief explanations. First, the calculation for Process A could have been accomplished with either form of the equation, since Process A was perfectly centered and equidistant from both specification limits. The C_{pk} for Process B is a negative value—this will occur whenever the process average is beyond a specification limit, either low or high. When the C_{pk} is negative, more than 50% of the product is out of specification. Finally, the C_{pk} for Process C is less than 1—this indicates some portion of the product manufactured in this process is beyond the specification limit. (That is, less than 99.73% of the product meets the specification limits.) The C_{pk} calculation tells us more about the process than the C_p calculation, because process centering is a part of the C_{pk} equation.

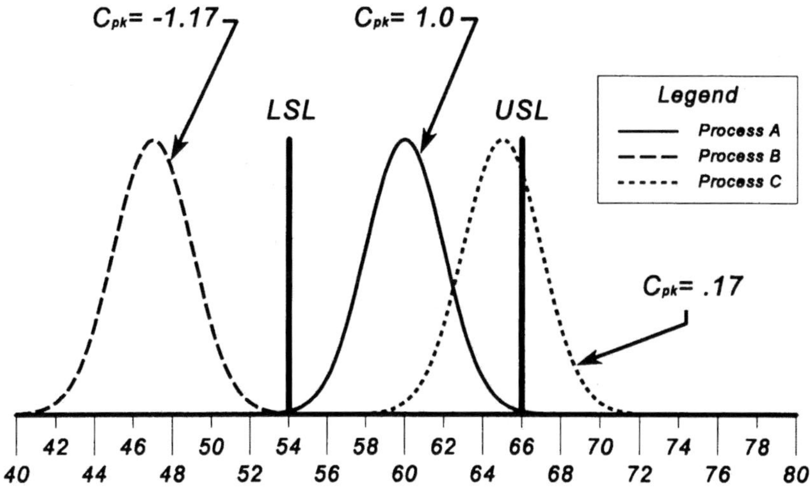

Figure 6-12: Various Distributions with C_{pk} Values Shown

This has been a necessarily brief summary of SPC principles. The references for this chapter contain a wealth of information on statistical techniques for process control and improvement. For our purposes, we shall examine the application of $\bar{x}$ - R and I-MR charts in complex manufacturing processes in the following chapter.

References

1. American Society for Testing and Materials, *Manual on Presentation of Data and Control Chart Analysis (MNL 7),* 6th edition, ASTM, Philadelphia, 1995.

2. Automotive Industry Action Group, *Fundamental Statistical Process Control Reference Manual,* AIAG, Troy MI, 1991.

3. Grant, E. L., Leavenworth, R. S. *Statistical Quality Control,* 5th edition, McGraw-Hill, New York, 1980.

4. Shewhart, W. A. *Economic Control of Quality of Manufactured Product,* Van Nostrand, New York, 1931; republished by ASQC Quality Press, Milwaukee, 1980.

7

Complex Process Control

Chapter 6 provided a review of SPC techniques. Here, we shall briefly examine how SPC can be affected by process components with multiple stations. We will also discuss how to apply SPC to such a process. The first part of this text illustrates how process designs can induce variation and complexity. As we witnessed, without careful design this variation can be so large that a process may not repeat a manufacturing path for many years. We also witnessed that, even if we apply the techniques discussed in Chapter 3, we may still have complex processes with a large number of paths. We might wonder how we can apply meaningful controls to such processes, and whether or not it can be accomplished in a cost-effective manner.

Complex Process Distributions

Consider the process component represented by Figure 7-1. It is a 6 station rotating turret. If we desire to monitor this process, one possible approach might be to develop and maintain an $\bar{x} - R$ chart for the output characteristic or characteristics determined by this component. This is a

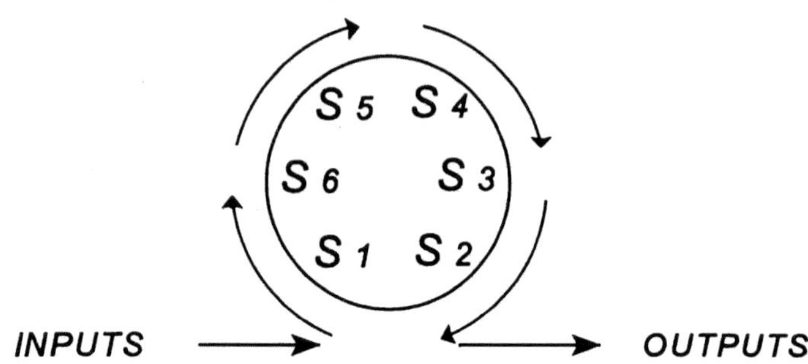

Figure 7-1: Turret Operation

fairly standard approach in many manufacturing facilities. The control chart might be established on the basis of a sample size equal to five, with the frequency of checking determined by factors such as: criticality of the characteristic being monitored, demonstrated process capability, the line speed, and time required to perform the necessary measurements.

Example 7.1

The control chart illustrated in Figure 7-2 is based upon actual data from a manufacturing process component with six stations. The engineering specification limits for the overall length of the part in this example are .445 ± .010". The SPC software used to produce the $\bar{x} - R$ chart indicates the data is normally distributed, with an average of .4455 inches and a standard deviation of .0020 inches. Both the average and range portions of the control chart indicate the process is in control, and therefore, stable. The software package also predicts a "true normal" distribution of individuals, as shown in Figure 7-3.

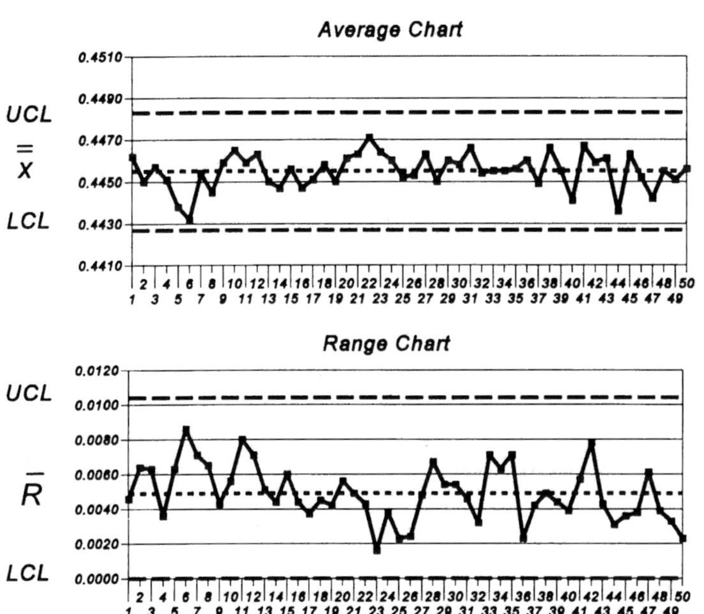

Figure 7-2: Control Chart for Six Stations, Sample Size of 5

The capability, or C_p of the process is:

$$C_p = \frac{tolerance}{6\ \hat{\sigma}_{ind}} = \frac{.455 - .435}{6 \times .002} = \frac{.020}{.012} = 1.67 \qquad (7\text{-}1)$$

The C_{pk} of the process is calculated with the specification limit nearest to the process mean, as shown in Equation 7-2:

$$C_{pk} = \frac{USL - \hat{\mu}}{3\ \hat{\sigma}_{ind}} = \frac{.455 - .4455}{3 \times .002} = \frac{.0095}{.006} = 1.58 \qquad (7\text{-}2)$$

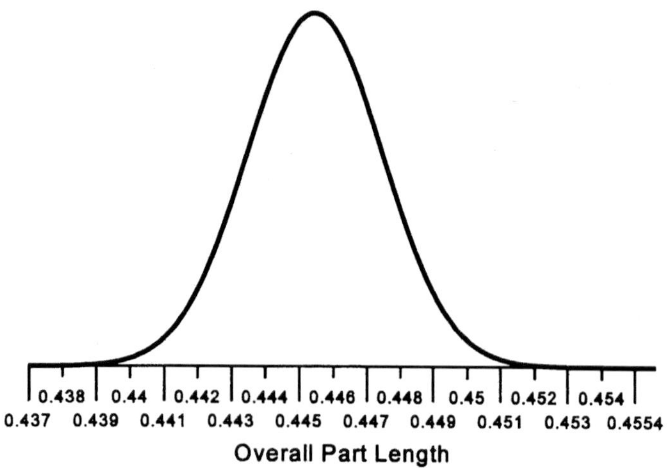

Figure 7-3: Combined Distribution of All Stations

This approach to process control might seem logical, however, it overlooks a basic tenet of our earlier discussions. As we examined in the first part of this book, the stations on this rotating turret will not produce *exactly* the same part. Even if they were so nearly identical in new condition that our measurement devices could not determine any difference between the stations, they still could not be reasonably expected to all wear at the exact same rate and in the exact same fashion. We are certain to see differences in the outputs produced by these stations. To determine how large these differences are, we need to isolate parts from each station and determine their respective measurements. We must then calculate their average and standard deviation, and produce the normal curves corresponding to the output from each station. The distributions corresponding to each of the six stations in this example are shown in Figure 7-4.

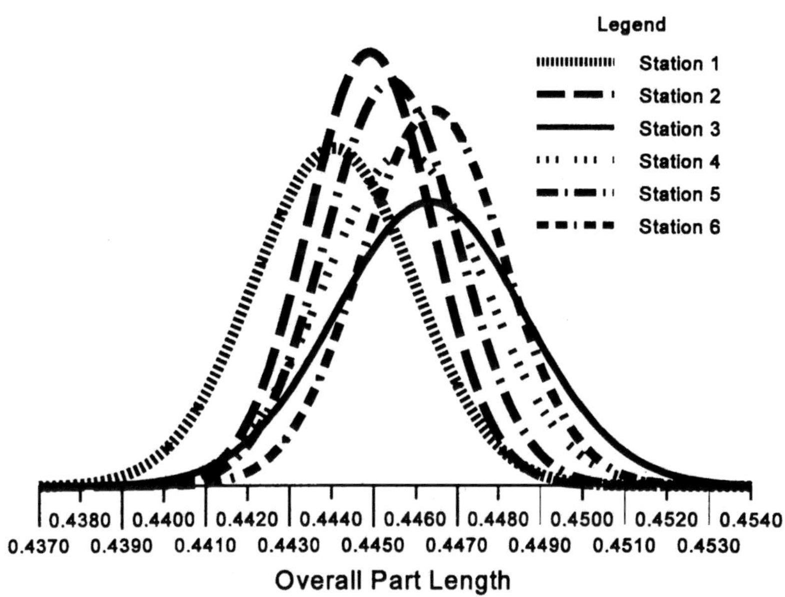

Figure 7-4: Distributions by Station

We stated in the previous chapter that the total area under a normal curve is always equal to 1. In Figure 7-4, we can see the normal curve representing output from Station 3 is shorter in height than the other five curves. Since Station 3 is shortest in height, it must have the widest distribution to be of equal area with all of the other curves. We know, therefore, from our quick visual scan that Station 3 exhibits the largest amount of variation. Calculation of the process capability, or C_p, should be accomplished based upon the output of Station 3. This station has an average of .4463 inches and a standard deviation of .0023 inches.

$$C_p = \frac{tolerance}{6 \, \hat{\sigma}_{ind}} = \frac{.455 - .435}{6 \times .0023} = \frac{.020}{.0138} = 1.45 \qquad (7\text{-}3)$$

The true process capability is not 1.67, as calculated from the combined distribution in Equation 7-1. It is 1.45, based upon the output drawn from Station 3. This difference may seem slight, but it is important when we consider that some customers require a minimum C_p of 1.67 for each manufacturing process.

We can also see that Station 3 produces the parts closest to the upper specification limit, and Station 1 produces parts closest to the lower specification limit. If we want to calculate the C_{pk} of the process, we must first compare the output of Station 3 to the upper specification limit. This will determine the smallest C_{pk} value with respect to the upper specification limit, since Station 3's output is closest and exhibits the largest amount of variation. This calculation is shown below.

$$Station\ 3,\ upper\ C_{pk} = \frac{USL - \hat{\mu}}{3\,\hat{\sigma}_{ind}} = \frac{.455 - .4463}{3 \times .0023} = \frac{.0087}{.0069} = 1.26 \qquad (7\text{-}4)$$

The calculation of minimum C_{pk} value for the lower specification limit is a bit more involved. To accomplish this we must consider the distribution closest to the lower specification, and all other distributions displaying larger amounts of variation than the closest one. The smallest of all the calculated values is the C_{pk} for the process. The station closest to the lower specification limit is Station 1. The C_{pk} for this station is calculated in Equation 7-5.

$$Station\ 1:\ C_{pk} = \frac{\hat{\mu} - LSL}{3\,\hat{\sigma}_{ind}} = \frac{.4440 - .435}{3 \times .0019} = \frac{.0090}{.0057} = 1.58 \qquad (7\text{-}5)$$

The normal curve for Station 4 has a standard deviation of .0019 inches, equal to Station 1. Since its standard deviation is the same es Station 1, we do not have to calculate its C_{pk} for the lower specification limit. Station 3's average and standard deviation were shown above. Calculating the lower specification limit C_{pk} for Station 3, we obtain:

$$Station\ 3,\ lower\ C_{pk} = \frac{\hat{\mu} - LSL}{3\,\hat{\sigma}_{ind}} = \frac{.4463 - .435}{3 \times .0023} = \frac{.0113}{.0069} = 1.64 \qquad (7\text{-}6)$$

For the combined distribution of all six stations, we had calculated a C_{pk} of 1.58 in Equation 7-2. We now see, from the series of calculations just performed, that the true C_{pk} of the process is 1.26, calculated from the relationship between the upper specification limit and the distribution of product output from Station 3.

Applying the $\bar{x} - R$ chart alone to this 6 station rotating component is misleading. The normal distribution of individuals calculated from the data of the $\bar{x} - R$ chart assumes a single population, with a single average and standard deviation. The true distribution, is really a compilation of output data for all six stations of the manufacturing component, as illustrated in Figure 7-5.

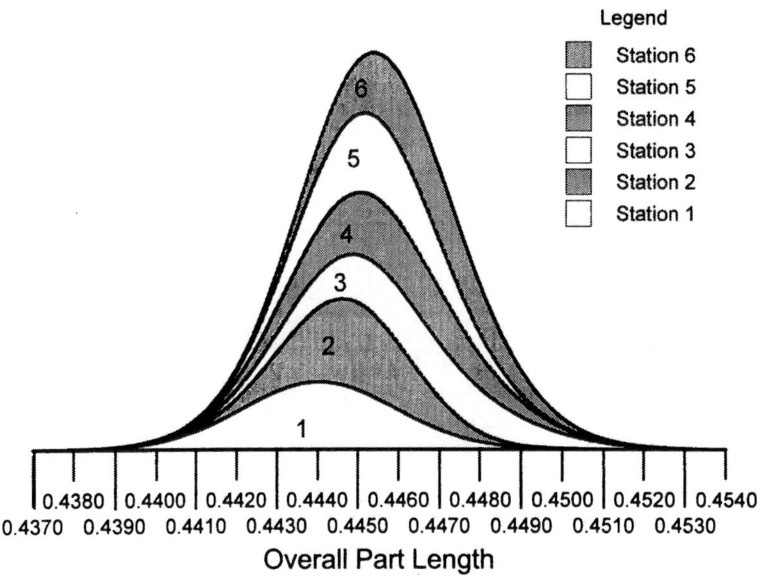

Figure 7-5: Station Contribution to Combined Distribution

Since each station of the manufacturing component is an independent entity of the process, it is more precise to identify and control them independently. After all, any individual station may exhibit signs of wear or improper setup at any time during the manufacturing process.

Albeit time-consuming and expensive, one alternative to the single $\bar{x}$ - R chart, would be to produce an $\bar{x}$ - R chart for each station. For illustrative purposes only, and with only 10 subgroups shown per station instead of the normal 25, these charts are all shown in the figure below.

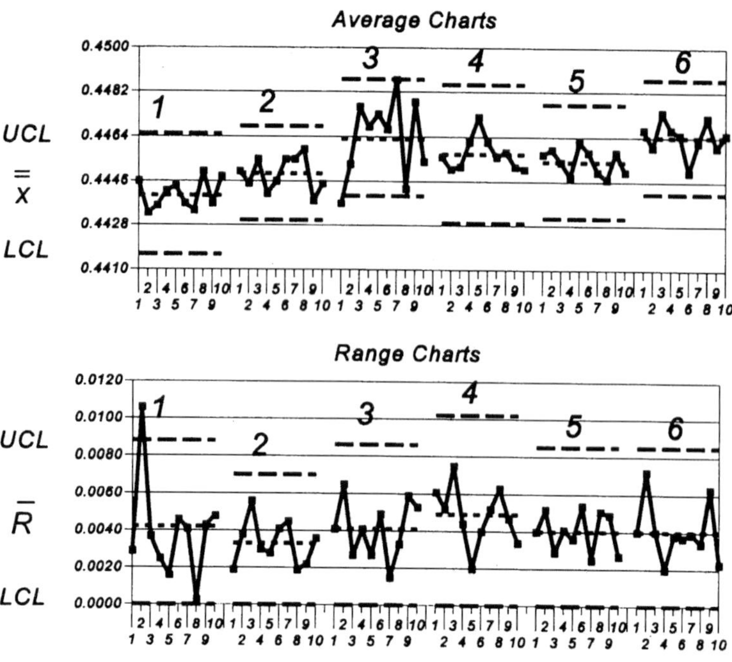

Figure 7-6: Control Chart with Limits for Each Station

From the individual station control charts, it is obvious that both Stations 1 and 3 are both out of control. This means that both are unstable and therefore, unpredictable. The C_p and C_{pk} values calculated for these two stations are not valid. Also equally important are that the other four stations are in control. This may indicate that the assignable cause or causes of variation in Station 1 and Station 3 are unrelated to material inputs to the process. The reason for the instability probably lies in the station setup or configuration. Clues such as this can be invaluable in troubleshooting and/or improvement efforts.

From this example, we can see that control charts for each individual station are more useful than one "composite" chart. This is true for calculating process capability, and for monitoring and improving process performance. The obvious negative aspect of charting each station is the increased amount of inspection and documentation time. Typically, the time required for complex processes is cost prohibitive. If we hope to be successful in a production environment, we must identify less expensive methods to monitor and control complex processes.

A second alternative to the single $\bar{x}$ - R chart would be the construction of individuals and moving range charts (I-MR) for each station. For Example 7.1, the I-MR chart for Station 1 is shown below.

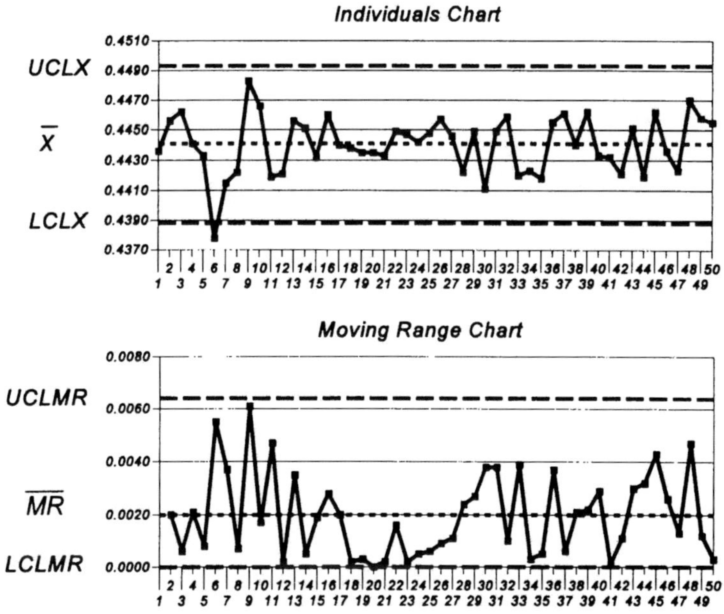

Figure 7-7: I-MR Control Chart for Station 1

The I-MR control chart for Station 1 clearly indicates the product from this station is out of control. Had we known this, we would not have attempted to calculate process capability from this data. Rather, we would have searched for the assignable cause of this variation, and eliminated it from our process. The I-MR chart, although not as sensitive as the $\bar{x}$ - R chart, would have provided useful insight into the state of control for Station 1. What about the other station that was unstable, namely Station 3? The I-MR control chart for this station is shown in Figure 7-8, below.

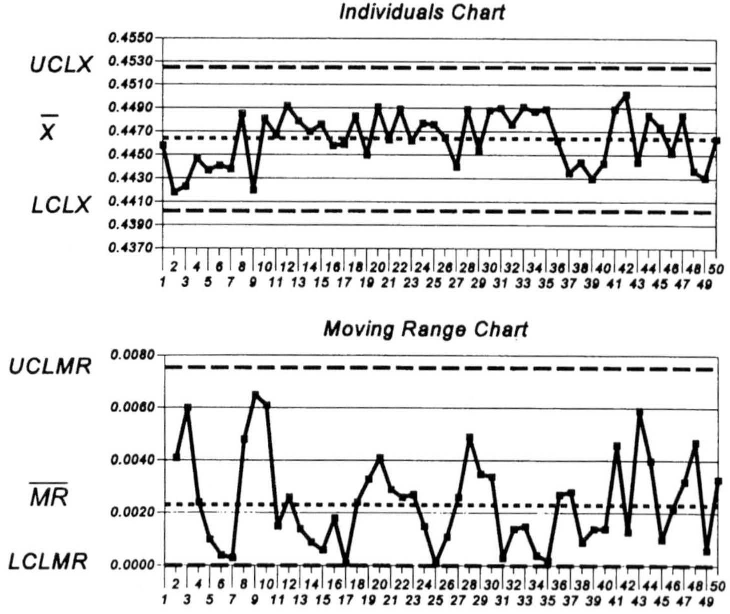

Figure 7-8: I-MR Control Chart for Station 3

This chart also demonstrates an out of control condition. Although there are no points evident beyond the control limits, the first seven points on the individuals chart are all below the average. This is an indication that the process may have been affected by some assignable cause of variation which caused the process average to become larger after the first seven pieces sampled. With respect to both stations, the I-MR chart indicates the presence of assignable cause variation. These charts can be important process control and monitoring tools for parallel components and complex processes. The principal drawback from a cost perspective is the need to keep one chart for every station. While this might be acceptable for a component with only six stations, it would still be prohibitive for a process with many stations. In the food processing industry, for example, it is not unusual for rotational components in filling applications to have 40 or 50 stations. We must still devise a plan for monitoring complex processes that is realistic, while at the same time providing the information we need for control and improvement efforts.

Monitoring Complex Processes

Both the $\bar{x} - R$ chart and the I-MR chart examined in the previous section offer distinct advantages with respect to process control. The $\bar{x} - R$ chart is almost universally understood, relatively easy to complete, and provides a simple method to compute process capability indices. It provides us with information such as the overall average process output that is unavailable from I-MR charts, unless of course we calculate the overall average manually. On the other hand, it is not capable of providing information critical to process control when dealing with multiple paths. It will not uniquely identify which station or stations are unstable. In addition to the lack of information, the capability indices calculated on the basis of the $\bar{x} - R$ chart may not be valid for the process being monitored.

The I-MR chart, while excellent for monitoring the output from individual stations, is not as sensitive as the $\bar{x} - R$ chart. A change in input material which affects all stations of the complex process would probably be detected more quickly by an $\bar{x} - R$ chart. Also, unless we are utilizing computer software for our SPC system in manufacturing, completion of I-MR charts requires more work. More ranges need to be calculated and more points need to be plotted. There are simply more charts to create, maintain, analyze, and file.

For Example 7.1, where we were dealing with a six-station component, one method to control this process would require adding an extra sample to the control chart and recording individual values by station number. This would allow us to produce either type of control chart. For instance, the sample columns on the control chart would appear as illustrated in the following table:

Table 7-1: Data Collection for Six-Station Component

DATE / TIME								
S T A T I O N	1	⇑	⇐ I-MR chart station #1 ⇒					
	2	$\bar{x} - R$	⇐ I-MR chart station #2 ⇒					
	3	chart	⇐ I-MR chart station #3 ⇒					
	4	all	⇐ I-MR chart station #4 ⇒					
	5	stations	⇐ I-MR chart station #5 ⇒					
	6	⇓	⇐ I-MR chart station #6 ⇒					
SUM								
AVERAGE								
RANGE								

This arrangement requires the operator to identify which station the SPC sample is obtained from and to record the measured characteristic in the appropriate row. The $\bar{x} - R$ chart above is completed with a sample size of six instead of five. The I-MR charts would be completed at the start of the run to verify process stability. They could then be updated intermittently thereafter, as required by the nature of the process and the criticality of the application. Of course, with appropriate computer software, the $\bar{x} - R$ chart and the I-MR charts could all be completed at once after each set of data is entered. This would enable us to examine all of the characteristics of the process simultaneously.

This approach offers a number of advantages over the standard $\bar{x}$ - R chart approach. First of all, it gives us the option of producing an Individuals and Moving Range Chart (I-MR) for each station of the component. This information can be invaluable in maintaining the process, with respect to both product integrity and scheduling maintenance on stations demonstrating instability due to wear of machine components. It also verifies the setup of each station on the turret. Even if we choose not to complete the I-MR charts in-process, having the data in this format is still important because it can save enormous amounts of time should a problem become apparent through $\bar{x}$ - R charting. Should the process go out of control, the operator could refer to the entries already on the control chart in search of the assignable cause of variation. Specifically, the data will indicate whether the problem is caused by:

1) the output of one of the stations changing, or
2) the entire process becoming erratic

For instance, if one of the stations is damaged or worn as discussed, one would expect to see a trend in the I-MR chart of the affected station. On the other hand, for a rotating turret, the output of all stations would demonstrate instability if a mechanical component such as the main journal bearing were to wear. In this instance, a trend would be visible in the $\bar{x}$ - R chart and in all or most of the I-MR charts.

While our six station example may be palatable to most, many operations contain rotational or parallel components with 30, 40, or more stations. While it is not feasible to record all of the stations and take samples from every one for routine SPC checks, a variation of this approach may still be applied. Under these conditions, each SPC check is pulled at random from the parallel components, with the individual measurement and the source of the part recorded in the SPC data table. To accomplish this, we need to add more columns to our data table, which function to record the station number the part was drawn from. This information is then used to produce I-MR charts. Sample SPC data is shown in Table 7-2.

Table 7-2: Data Collection for Complex Process Control

DATE / TIME							
		DATA	SOURCE	DATA	SOURCE	DATA	SOURCE
S A M P L E	1		(I-MR		(I-MR		(I-MR
	2		station		station		station
	3		numbers		numbers		numbers
	4		recorded		recorded		recorded
	5		here)		here)		here)
SUM							
AVERAGE							
RANGE							

Over time, with random sampling, the distribution of each station can be determined by isolating and recording measurements taken from each individual station. As was the case with the six-station example, it may not be desirable to perform this tracking on a continuous basis due to cost considerations unless we have computer software available to produce the charts automatically. Once again, the information can be invaluable when investigating problems, determining machine wear, or engaging in process optimization efforts.

References

1. American Society for Testing and Materials, *Manual on Presentation of Data and Control Chart Analysis (MNL 7),* 6th edition, ASTM, Philadelphia, 1995.

2. Automotive Industry Action Group, *Fundamental Statistical Process Control Reference Manual,* AIAG, Troy MI, 1991.

3. Grant, E. L., Leavenworth, R. S. *Statistical Quality Control*, 5th edition, McGraw-Hill, New York, 1980.

4. Shewhart, W. A. *Economic Control of Quality of Manufactured Product,* Van Nostrand, New York, 1931; republished by ASQC Quality Press, Milwaukee, 1980.

Part 3

Design of Experiments

8

Inputs and Outputs

The first part of this book dealt with the manufacturing process as a simple or complex group of components all combined to manufacture a product. In the second part, we examined simple but powerful techniques to identify problems and improve a manufacturing process. This section differs because we will emphasize not the components or their arrangement, but the inputs to the individual components and the resulting measurable outputs.

In Chapter 1, we defined a process as "one or more actions on inputs designed to produce a product with one or more specific output characteristics." We then proceeded to examine and classify various process layouts and study the variation which might result. Our efforts centered on the variation which might be added to the inputs by the arrangement of process components. For our purposes, process inputs are defined as follows:

> *Any factor which will influence the measurable characteristics of the process outputs is considered an input to the process.*

There are two categories of process inputs: control factors and noise factors.

Control Factors

Control factors, are those factors we establish, monitor, and *control* within prescribed limits in order to influence or determine the output characteristics of a process. Designed experimentation involves the planned and documented study of control factors. The goal of the experiment is to assess the effects the control factors have on one or more characteristics of the output.

Control factors can be either continuous or discrete variables. A typical continuous variable will be specified as a target ± some allowable tolerance. For example, if we are dealing with a mixing process, the weight of a specific ingredient would be a continuous input specified as a control factor. The weight specification might be stated as 150 ± 10 lb Any level of addition between 140 lb and 160 lb would be possible and acceptable (although perhaps not preferred). If the measurement and addition of the ingredient were a stable process, with a C_{pk} of approximately 1, the variation in the input might appear as illustrated in Figure 8-1.

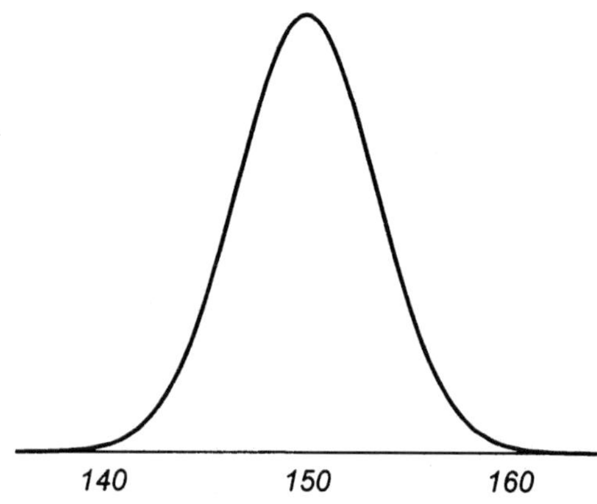

Figure 8-1: Continuous Control Factor

A discrete variable as an input factor can assume only fixed values. For example, a process may require a specific type of resin as part of the ingredient mix. The purchase specification may allow three sources, suppliers A, B, and C. In this case the input is controlled, however, there are three and only three levels considered as inputs to the manufacturing process. (We are not considering the variation inherent in each respective supplier's product, since it is beyond our control and might be proprietary.) Figure 8-2 illustrates this discrete input with three possible values or levels.

Figure 8-2: Discrete Control Factor Levels

A process is typically affected by many control factors concurrently. Whether the control factor is continuous or discrete, the convention for representing them in our process diagrams is as follows:

Figure 8-3: Control Factors

Noise Factors

The other category of input factors we need to examine are noise factors.[2] This category, from Dr. Genichi Taguchi's approach to experimentation, refers to inputs which influence process characteristic outputs, but are either unknown, or too difficult or expensive to control. They are often overlooked when examining process performance or attempting process improvement through experimentation.

Noise factors can be divided into two categories: inner noise and outer noise. Inner noise refers to factors internal to the process or product being manufactured such as wear on a machine shaft or bearing. Outer noise refers to factors outside of the process or manufactured product which affect it. Examples of outer noise are ambient temperature, pressure, and humidity. The convention for representing noise factors in our process diagrams is as follows:

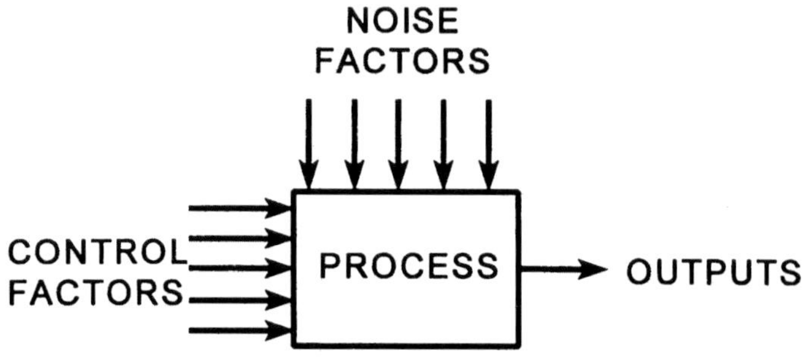

Figure 8-4: Noise Factors

---

[2] ©Copyright, American Supplier Institute, Inc., Allen Park, Michigan (U.S.A.) "Reproduced by permission under License No. 970101."

Outputs

An output is defined as the end product or service which results from subjecting inputs to a process. A process output is described in terms of quality characteristics which quantify the output's attributes. For example, if we are manufacturing a plastic molded part we might describe the process output in terms of color, length, width, and depth.

A quality characteristic is defined as a quantifiable attribute of the process output.

A good quality characteristic is an attribute measured by counting or gaging. Characteristics which are rated subjectively (such as visual quality ratings of good, fair, or poor) are generally not considered good quality characteristics to describe a process. This is due to the variation inherent in subjective ratings. Typically, a single physical output will have possess characteristics to describe its state or condition.

Quality characteristics can be either continuous or discrete. Continuous characteristics are attributes such as length, width, diameter, depth, etc. Discrete characteristics usually involve classifications such as pass/fail, or go/no go, etc.

Process failures are most often described in terms of select characteristics not adhering to manufacturing specifications. The purpose of designed experimentation in most manufacturing environments is either to meet specifications, or provide a superior product with respect to industry standards. In this text we will concentrate on three Taguchi strategies for improving quality characteristic outputs. These strategies are larger-the-better, smaller-the-better, and nominal-the-best.[3]

Larger-the-Better

Many quality characteristics are more acceptable as their average value increases. For example, consider a product where life is a critical characteristic, measured in hours. If the industry standard for this material

---

[3] ©Copyright, American Supplier Institute, Inc., Allen Park, Michigan (U.S.A.) "Reproduced by permission under License No. 970101."

is 1000 hours, the manufacturer could improve their competitive position and capture more market share by producing a material with an average life of 1200 hours as illustrated in Figure 8-5.

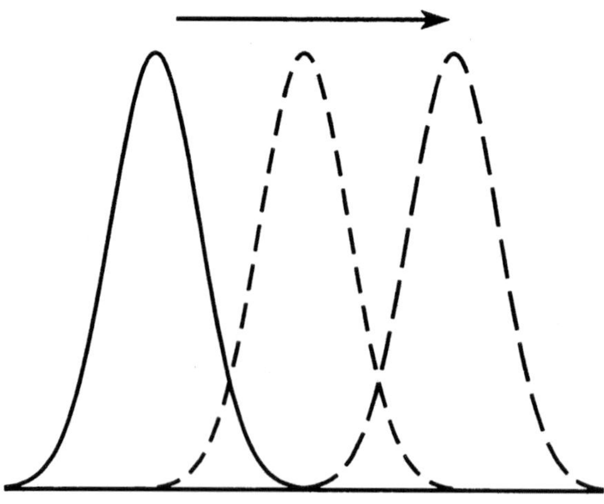

Figure 8-5: Larger-the-Better

A word of caution with respect to larger-the-better characteristics—yield is not generally a good quality characteristic for experimentation. Yield, or the lack thereof, usually results from multiple quality characteristics which fail to meet manufacturing specifications. In lieu of measuring process yield, consider measuring and tabulating all of the quality characteristic outputs which determine process yield.

Smaller-the-Better

Other quality characteristics are more acceptable as their average value decreases. For example, consider a characteristic such as surface blemishes in a painting process. If we count the number of occurrences per finished part, we naturally would desire the count to be zero. A part with zero blemishes is a high quality product.

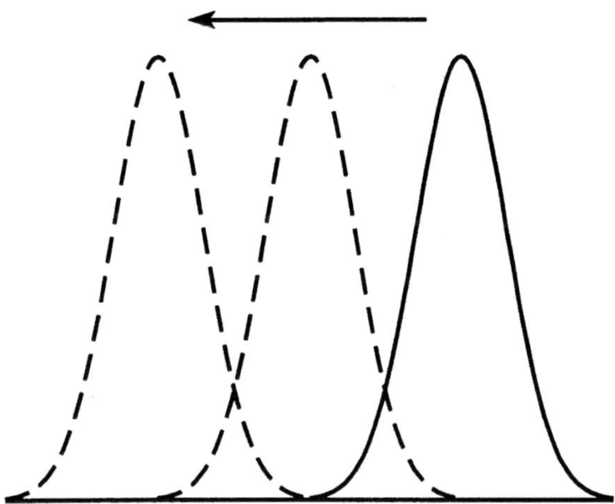

Figure 8-6: Smaller-the-Better

Nominal-the-Best

The final category of quality characteristic examined in this text deals with characteristics that require reduced variation about some nominal value. These characteristics are frequently specified on engineering drawings and include items such as: length, width, height, and all other dimensional characteristics with bilateral tolerances. While we often have a great deal of data collected on nominal-the-best characteristics in the form of inspection records, control charts, C_p's, and C_{pk}'s, they are most often not the subject of experimentation and improvement efforts in a typical organization. Quite frankly, this is caused by a lack of understanding and patience on the part of many manufacturing personnel. While it is easy to determine which control factors affect targeting, it is often more time consuming to determine which factors control the amount of variation in output characteristics. A nominal-the-best characteristic is depicted in Figure 8-7.

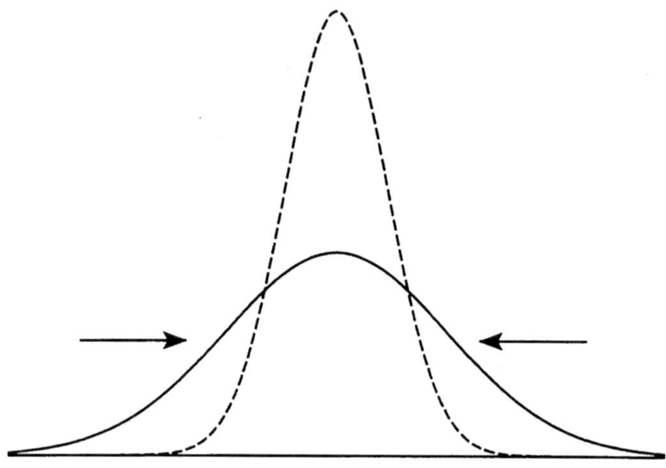

Figure 8-7: Nominal-the-Best

In practice, we often find quality characteristic responses that require both reduced variation and improved targeting with respect to some nominal value. This text will cover techniques which provide the ability to fulfill both requirements.

References

1. American Supplier Institute *Taguchi Methods, Introduction to Quality Engineering Course Manual,* Version 2.1, American Supplier Institute, Dearborn, MI, 1991.

2. Barker, T. B. *Engineering Quality by Design—Interpreting the Taguchi Approach*, Marcel Dekker, New York, 1990.

3. Barker, T. B. *Quality by Experimental Design,* 2nd edition, Marcel Dekker, New York, 1994.

4. Berger, R.W (ed.) and Pyzdek, T. (ed.) *Quality Engineering Handbook*, ASQC Quality Press, Milwaukee, WI, and Marcel Dekker, New York, 1992.

5. Lochner, R.H. and Matar, J.E. *Designing for Quality—An Introduction to the Best of Taguchi and Western Methods of Statistical Experimental Design*, Quality Resources, New York and ASQC Quality Press, 1990.

6. Montgomery, D.C. and Myers, R.H. *Response Surface Methodology —Process and Product Optimization Using Designed Experiments,* John Wiley & Sons, New York, 1995.

7. Taguchi, G. *System of Experimental Design,* UNIPUB/Kraus International Publications, White Plains, NY, and American Supplier Institute, Dearborn, MI, 1987.

8. Taguchi, G. *Introduction to Quality Engineering*, Asian Productivity Organization, Tokyo, 1986.

9

Control Factors

In the previous chapter we indicated that designed experimentation involves the planned and documented study of control factors, with the goal of assessing the effect control factors have on one or more characteristics of the output. This is exactly what most manufacturers do whenever they desire to increase their understanding or control of a process. However, without experimental design, this experimentation is typically accomplished by altering one factor at a time and attempting to assess its impact on the process output. This approach does not constitute an experiment. It is a simple test of one factor's influence on the process output, typically under inadequately controlled conditions.

The one-factor-at-a-time approach, although almost universally used, has at least two major flaws. First, emphasis is placed on controlling the factor being studied. This often allows other factors to vary within their normal operating range. At the conclusion of the test it is not possible to sort out or quantify the magnitude of influence these other factors may have had on the process output. The second shortcoming of the one-factor-at-a-time approach is its inherent inability to assess whether or not the factor being evaluated interacts with any of the other inputs to the process. As we will see, the experimental design techniques we are discussing in this and subsequent chapters will address both of these shortcomings.

The first step in the development of an experimental design is to select the control factors we wish to evaluate. The best approach for factor selection is through a brainstorming process, and the use of an Ishikawa Diagram. Once the factors have been selected, the next step is the establishment of factor levels.

Factor Levels

Consider the continuous control factor illustrated in Figure 9-1. If the input specification is from 140 lb to 160 lb there are an infinite number of possible values, limited only by the resolution of our measuring equipment. The specification tolerance of 150 ± 10 lb indicates any of these values would be allowed as process inputs.

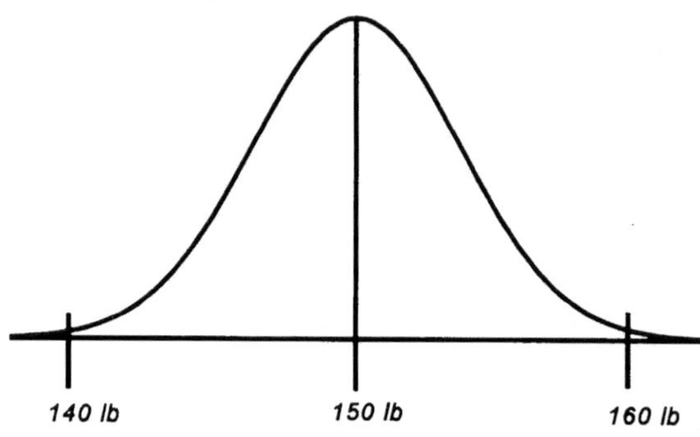

Figure 9-1: Continuous Control Factor

If we are attempting to maximize a characteristic output, we need to determine the effect different values of the input have on the output. For instance, what effect would an ingredient weight of 160 lb have? What about 140 lb, or perhaps 145 lb? If we want to explore weights beyond the current input specification, what about 170 lb and 130 lb?

In designed experimentation, we select different levels for the control factors, subject them to the process being studied, and quantify the magnitude of their effects on the quality characteristics we are attempting to control. This selection and notation of factor levels can sometimes be a source of confusion, mainly due to the different styles of notation found in current literature.

In classical, or Western experimental designs, factor levels are normally indicated by a +1 and a -1. This notation merely indicates different levels of some control factor input. For example, the Western notation could be used to correspond with level above and below the target value as illustrated in Figure 9-2.

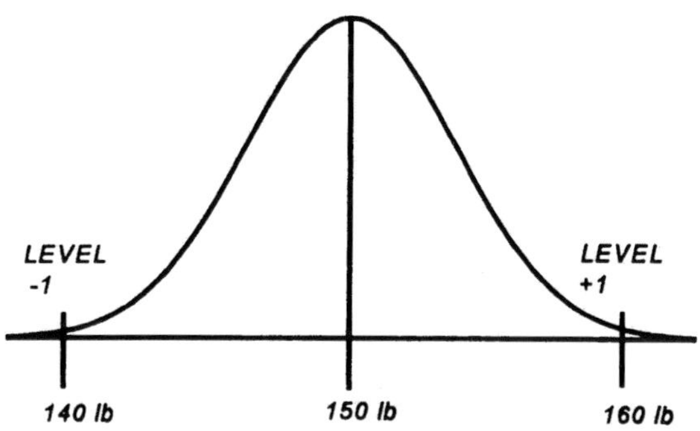

Figure 9-2: Selecting Factor Levels

While this selection of factor levels and the +1, -1 notation is certainly valid, it is not the only method we might employ. It would be equally correct to select factor levels and label them as illustrated in Figure 9-3. In this example, the designations of +1 and -1 are reversed. As we shall examine in later chapters, the notation will ultimately have an effect on the graphs we produce to analyze the experiment. It will have no effect, however, on the physical relationships between the inputs and outputs of the process.

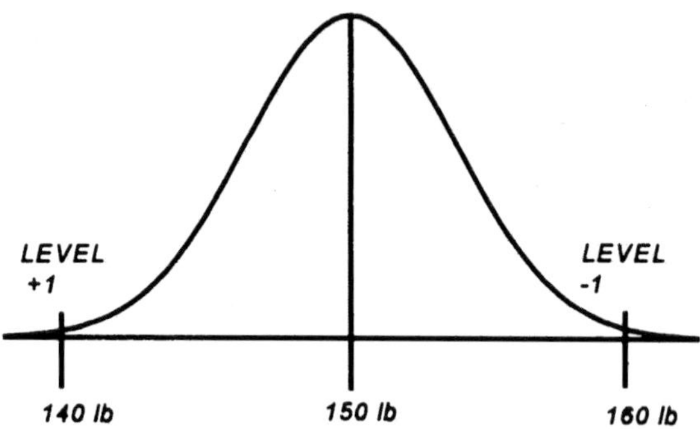

Figure 9-3: Selecting Factor Levels

In practice, the +1, -1 notation can be used to define any two factor levels, without regard to specification, tolerance, or past practice. Suppose we felt confident that ingredient mix weights below 150 lb were adversely affecting the quality characteristic we are trying to improve. We might choose to go beyond the current realm of experience and test a mix with 165 lb of ingredient, and compare the result with a standard mix of 150 lb. These factor levels are shown in Figure 9-4.

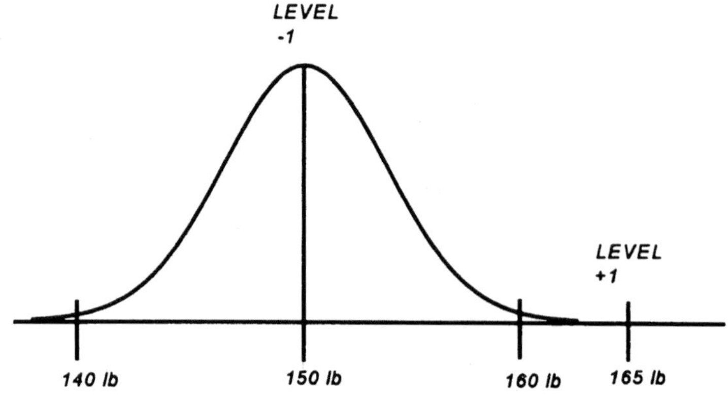

Figure 9-4: Selecting Factor Levels

Once again, the selection of values corresponding to the +1, -1 notation were arbitrary, at least in the sense that the level represented by the -1 does not have to be less then the level represented by the +1. It would also be proper to select 165 lb as the -1 level and 150 lb as the +1 level. In Taguchi experimental designs the factor levels are noted by 1, 2 notation. As with Western experimental designs, the 1, 2 labeling of factor levels is arbitrary—that is, the notation is used merely to differentiate between the two control factor levels. Occasionally, you might see some experiments with levels labeled simply "low" and "high."

Selection of +1, -1 levels for discrete control factors is also at the discretion of the designers of the experiment, and perhaps illustrates the concept more clearly. Consider the discrete control factor represented in Figure 9-5. Here we have three sources of supply for a specific type of resin used as an input to a process.

Figure 9-5: Discrete Control Factor Levels

If we were presently using resin from supplier A in our process and wanted to evaluate the effect of substituting resin from supplier B, we could assign resin A to level +1 and resin B to level -1. Of course, if we conduct experimentation to improve our processes with only one factor at a time, the +1, -1 notation adds no value to our efforts. The importance of this notation will become clear in the following chapters.

Factor Combinations

In order to examine factor combinations, we will switch to the 1, 2 notation used by Taguchi in lieu of the Western notation of +1, -1. Consider a process with two control factors, A and B. The process flow diagram is drawn in Figure 9-6.

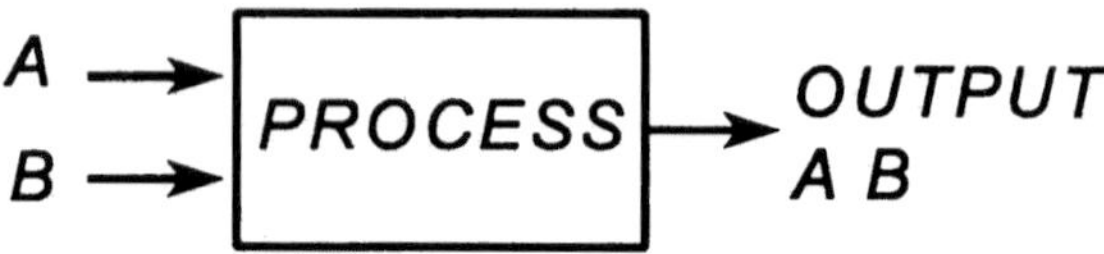

Figure 9-6: Two Control Factors

If we allow factor A to run at two different levels we would expect to get variation in the quality characteristics of our output. If we denote one level of control factor A as A_1, and the second level of control factor A as A_2, we could show both of the resulting process diagrams as illustrated in Figure 9-7. This dual representation indicates that the output produced when A is at level 1 is quite possibly different than the output when A is at level 2.

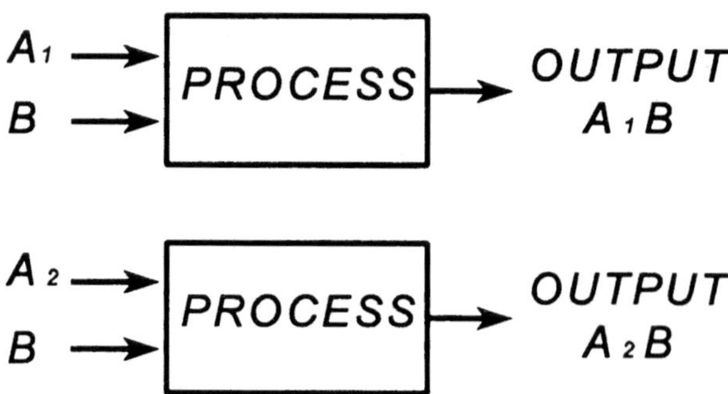

Figure 9-7: Two Control Factors, Factor A at Two Levels

If we also allow control factor B to vary between two levels, and use the same form of notation as we did for A, all resulting process flow diagrams are shown in Figure 9-8.

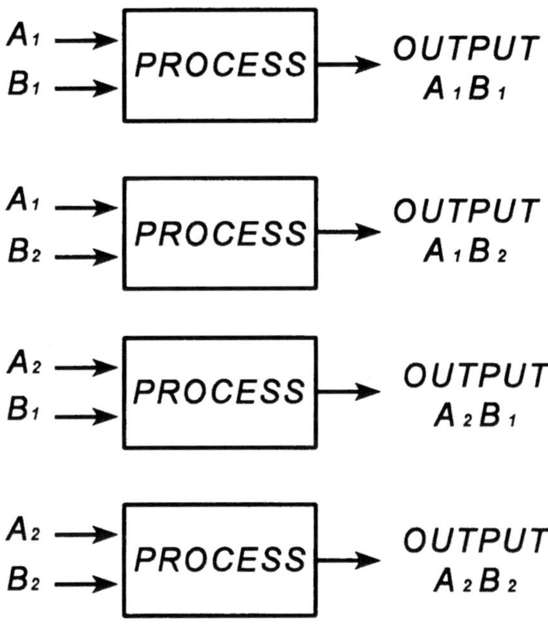

Figure 9-8: Two Control Factors at Two Levels

As the number of factors and levels increase the possible combinations increase exponentially. The form of this relationship is:

$$Possible\ Combinations = L^n$$

(9-1)

$$where\ L = number\ of\ levels$$

$$n = number\ of\ factors$$

Although we could continue representing combinations with multiple process flow diagrams, this is time consuming and unnecessary. An easier method is through the use of a tree diagram. This allows us to see the possible combinations of input factors and is very easy to complete. Consider once again two control factors at two levels. All possible combinations of these factors are shown below.

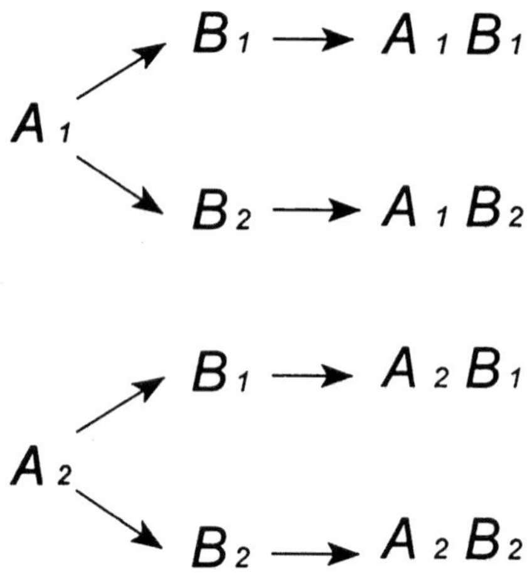

Figure 9-9: Diagram for Two Factors at Two Levels

The tree diagram illustrates the four possible combinations for the two control factors at two levels. You can verify this is correct by recalling Equation 9-1:

$$Possible\ Combinations\ =\ L^n = 2^2 = 4 \qquad (9\text{-}2)$$

Thus far, we've only examined two control factors at a time. Because of the exponential relationship shown above, the number of possible combinations increases very rapidly. Figure 9-10 illustrates three control factors, each set at two levels.

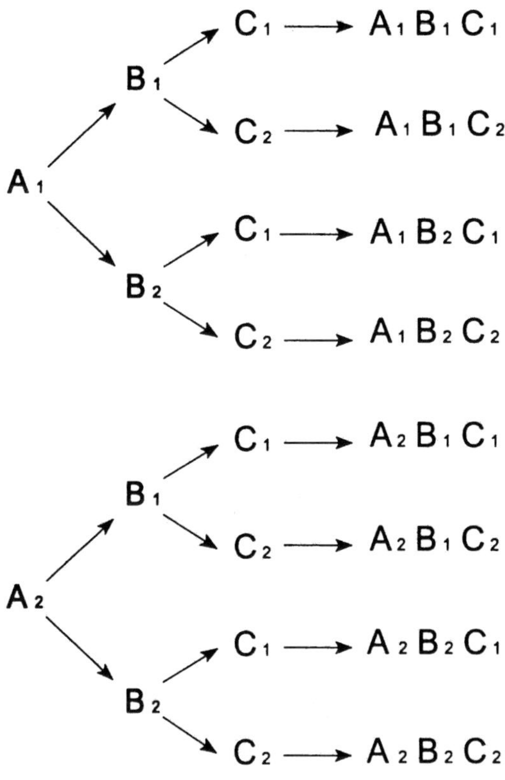

Figure 9-10: Three Factors at Two Levels

If we assume the control factors A, B, and C all have significant effects on the quality characteristics of the process output, we are diagraming 8 different combinations of inputs and 8 unique outputs.

$$Possible\ Combinations = L^n = 2^3 = 8 \qquad (9\text{-}3)$$

Most processes in today's complex manufacturing environment have many possible control factors. Whether or not all are important in controlling the quality characteristics of the output is often uncovered through the application of experimental design techniques. It's important to remember that control factors are typically selected because designers or users of the process feel they are important to insure uniformity and end-product integrity. When control factors can be proven unimportant with regard to product integrity, they are candidates for relaxed specifications and controls. Specifications which are unnecessarily prohibitive add to the cost of manufacture, without adding any real value to the product. Such specifications are a form of waste and are prime candidates for cost reduction efforts.

Control factors do not always have the same number of levels in experimentation. For instance, consider the control factors A, B, and C, where factors A and C are set at two levels, and factor B is set at three levels. The diagram for these factors is shown in Figure 9-11, on the following page. In this example, the number of possible combinations is calculated as follows:

$$\textit{Possible Combinations} = L_a^{n_a} \times L_b^{n_b} \times L_c^{n_c} \dots \tag{9-4}$$

$$= 2^2 \times 3^1 = 12$$

Experimental designs can be developed to address these conditions, however, they are beyond the scope of this text. Typically, we perform experiments with three-level designs when we suspect that quality characteristics respond to changes in control factor levels in a nonlinear fashion. We will concentrate on two-level experimental designs in this text. However, it is important to recognize that they are not the only type of design possible. For most applications, the two-level designs are more than adequate for process optimization efforts. These designs are easy to design, perform, and analyze. They will identify which factors actually control the quality characteristics of the process outputs.

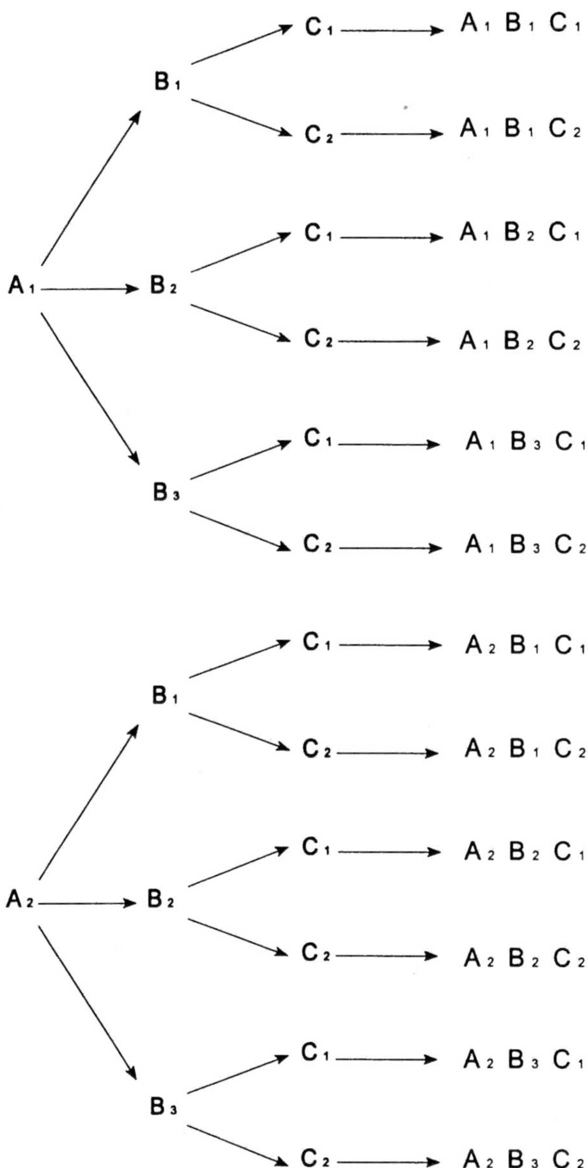

Figure 9-11: Two Factors at Two Levels, One Factor at Three Levels

References

1. American Supplier Institute *Taguchi Methods, Introduction to Quality Engineering Course Manual*, Version 2.1, American Supplier Institute, Dearborn, MI, 1991.

2. Barker, T. B. *Engineering Quality by Design—Interpreting the Taguchi Approach*, Marcel Dekker, New York, 1990.

3. Barker, T. B. *Quality by Experimental Design*, 2nd edition, Marcel Dekker, New York, 1994.

4. Berger, R.W (ed.) and Pyzdek, T. (ed.) *Quality Engineering Handbook*, ASQC Quality Press, Milwaukee, WI, and Marcel Dekker, New York, 1992.

5. Lochner, R.H. and Matar, J.E. *Designing for Quality—An Introduction to the Best of Taguchi and Western Methods of Statistical Experimental Design*, Quality Resources, New York and ASQC Quality Press, 1990.

6. Montgomery, D.C. and Myers, R.H. *Response Surface Methodology —Process and Product Optimization Using Designed Experiments*, John Wiley & Sons, New York, 1995.

7. Taguchi, G. *System of Experimental Design*, UNIPUB/Kraus International Publications, White Plains, NY, and American Supplier Institute, Dearborn, MI, 1987.

8. Taguchi, G. *Introduction to Quality Engineering*, Asian Productivity Organization, Tokyo, 1986.

10

Experimental Arrays

For many individuals, experimental arrays are extremely confusing and often misunderstood. I am personally convinced this is a result of the methods typically employed in the presentation of the subject matter. That is, we are generally introduced to experimental arrays through the use of statistics and/or complex mathematics. While these methods are the very basis of the arrays, and are critical from a scientific perspective, they provide value only to those who understand them.

Our goal in this chapter is to develop the arrays in a fashion easily understood to everyone. Possessing a degree in statistics, mathematics, or engineering is *not* a prerequisite for the effective implementation of basic experimental design techniques in manufacturing environments. After all, most people have no idea of the thermodynamics of the internal combustion engine, yet they are quite proficient at driving automobiles.

In making this presentation, I am not discounting the disciplines supporting the formation of experimental arrays. For those interested in developing a deeper level of understanding, there exists a rather large body of work on the subject matter. You need only visit a college library or read the references at the end of this chapter.

Main Effects

Let's begin with two factors we suspect are important for controlling quality characteristics in a process output. For this example, the quality characteristic we want to measure is overall part length. Suppose we are dealing with a molding process and suspect both the amount and the source of resin in our formulation control this characteristic. We want to alter the levels of both factors and determine the impact on part length. Labeling these control factors as A and B, and selecting two levels for each provides enough information to begin constructing an experimental array as follows:

Control Factor A : resin weight, lb
    Level A_1 : 140 lb
    Level A_2 : 160 lb

Control Factor B : source of resin
    Level B_1 : supplier 1
    Level B_2 : supplier 2

Quality Characteristic : Length, inches

We begin building our experimental array as follows:

Trial	A	B	Response
1	1	1	R_1

The 1 in the Trial column indicates this is the first combination of control factor levels in the experiment. (The combination of control factor levels unique to a particular experimental trial is called the *treatment combination,* and is often abbreviated as *tc.*) The 1 in the A and B columns indicates both control factors are established at level 1 for this trial. The R_1 in the Response column is where the resulting value of the quality characteristic (part length) for Trial 1 is recorded. If the A_1 B_1 combination were actually run and a part length of 1.505" was measured as the output, the partial array would look as follows:

Trial	A	B	Response
1	1	1	R_1=1.505

From this first trial, we have learned the part length was 1.505" when the control factor combination was A_1 B_1. Our experimental goal is to find out how factors A and B *individually* effect part length. So far, we have only determined the resulting length when A is at level 1 and B is also at level 1. We need to know the average value we will obtain when A is at level 1, and B varies. That is, we need to know the average part length when 140 lb of resin is used in our mix, irrespective of whether the resin is from supplier 1 or from supplier 2.   The only way to accomplish this is to run another test with factor A at level 1 (140 lb of resin) and factor B at level 2 (i.e., with resin from supplier 2). The experimental array with both trials is shown below. For this example, assume the part length resulting from the Trial 2 combination of A_1 B_2 was 1.515 inches.

Trial	A	B	Response
1	1	1	R_1=1.505
2	1	2	R_2=1.515

Since A_1 has been tested across the entire experimental range of factor B, we can now calculate the average response when A is at level 1 as shown in Equation 10-1:

$$A_1 = \frac{R_1 + R_2}{2} = \frac{1.505 + 1.515}{2}$$

$$\text{(10-1)}$$

$$A_1 = 1.510$$

Now that we know the resulting average part length when our process is run with 140 lb of resin, we must determine the average part length when 160 lb of resin are used. Once that is determined, we can calculate the difference between the two average part lengths and determine the effect of altering the resin weight. The difference between two average level responses for a control factor is known as a *main effect*. It is the measure of how the quality characteristic response changes when the control factors are altered from one level to another.

We must now repeat the first two trials, with the resin weight controlled at 160 lb. Building our complete array one trial at a time, we now add a third trial, with factor A at level 2 and factor B at level 1. Our partial experimental array, with the length response filled in, now appears as follows:

Trial	A	B	Response
1	1	1	R_1=1.505
2	1	2	R_2=1.515
3	2	1	R_3=1.520

Trial 3 allows us to determine the response when the control factor combination is $A_2 B_1$. As we did with A at level 1, we need to run another trial to insure we have evaluated A_2 at both possible levels of B. Creating another trial for our experimental array to run the $A_2 B_2$ combination, and filling the experimental response in, we obtain:

Trial	A	B	Response
1	1	1	$R_1=1.505$
2	1	2	$R_2=1.515$
3	2	1	$R_3=1.520$
4	2	2	$R_4=1.525$

We have now run A_2 with both possible levels of control factor B and we calculate the average response when factor A is at level 2 as:

$$A_2 = \frac{R_3 + R_4}{2} = \frac{1.520 + 1.525}{2}$$

$$A_2 = 1.5225$$

(10-2)

To visualize this calculation more clearly, it helps to "block" out all rows and columns except the ones of interest for the calculation of average responses. (We will also drop the R_1-R_4 notation.) For example, to calculate the average response for A_1 we have:

Trial	A	B	Response
1	1	1	1.505
2	1	2	1.515
3	2	1	1.520
4	2	2	1.525

To calculate the average response for A_2 we block out all other rows and columns, and obtain the following:

Trial	A	B	Response
1	1	1	1.505
2	1	2	1.515
3	2	1	1.520
4	2	2	1.525

Our calculations are valid within the range of experimental settings because the average response for control factor A was determined by running A_1 at both levels of control factor B, and by running A_2 at both levels of factor B. At this point one might think that all we have accomplished in these tests is the evaluation of control factor A at two levels. The concept behind designed experimentation, and the real "power" of the experimental arrays we discussing is about to be exposed.

Let's re-examine our experimental array. We want to determine how the levels of control factor B are dispersed with respect to the two levels of control factor A. First, examine the levels of A when B is at level 1.

Trial	A	B	Response
1	1	1	1.505
2	1	2	1.515
3	2	1	1.520
4	2	2	1.525

Clearly, B_1 has been run with both possible levels of control factor A. We can therefore calculate an average response for control factor B at level 1, and we can be confident that the result will not be dependent upon any single level of A. Calculating the average response for B_1 we obtain the following:

Trial	A	B	Response
1	1	1	1.505
2	1	2	1.515
3	2	1	1.520
4	2	2	1.525

$$B_1 = \frac{1.505 + 1.520}{2}$$

$$B_1 = 1.5125$$

(10-3)

Next, we examine the experimental array when control factor B is at level 2, as illustrated below.

Trial	A	B	Response
1	1	1	1.505
2	1	2	1.515
3	2	1	1.520
4	2	2	1.525

For factor B at level 2, factor A has been run at both levels. We can therefore calculate an average response for B_2, and this result will not be dependent upon any single level of A. The shaded array, and the calculation of the average part length which results when control factor B is at level 2 (resin from supplier 2), is as follows:

Trial	A	B	Response
1	1	1	1.505
2	1	2	1.515
3	2	1	1.520
4	2	2	1.525

$$B_2 = \frac{1.515 + 1.525}{2}$$

$$B_2 = 1.520$$

(10-4)

Examining the experimental array for control factor A and B illustrates the concept of the arrays we are studying. For every level of factor A, factor B has been run at level 1 and at level 2. Likewise, for every level of factor B, factor A has been run at level 1 and at level 2. This allows us to calculate the average level effect on the process output for both control factors, after running four trials. The arrangement is often referred to as a balanced array. We shall explore this further in a later chapter.

This arrangement, as we shall see in the next chapter, will also enable us to determine and quantify the magnitude of interaction between factors A and B. But even if we did not have this additional ability to examine interactions, the experimental array represents a significant improvement over the one-factor-at-a-time approach. If the four trials had been accomplished as four one-factor-at-a-time tests, we would probably not have maintained adequate control over the factor not being immediately evaluated. For instance, if we were testing A at level 1, where would we have established the level of factor B? Would we have kept the same level of B when testing A_2? If we had maintained this type of control, the results we obtained for A at levels 1 and 2 would not have been valid, or averaged over both levels factor B. The results obtained for A would have been valid for one supplier's resin, but may not have reacted the same when applied to resin from the remaining supplier.

Response Table

We have now established the average level response for each of the two control factors in our example. Our next step is to calculate the main effects. One method to accomplish these calculations is through the use of a response table. The results of our sample experimental array are summarized in the response table which follows:

	Control Factor A	Control Factor B
Level 2 Average	1.5225	1.5200
Level 1 Average	1.5100	1.5125
Difference	.0125	.0075

Reviewing the bottom row of the response table reveals which control factor has the greatest influence or effect on the quality characteristic we have measured. The data in the response table above indicates a change in factor A from level 1 → level 2 causes a .0125" increase in length. Changing factor B from level 1 → level 2 results in a .0075" increase in length. Later in this text, we will explore methods to confirm these results, and insure they are not due to random variation or the result of extraneous inputs which we are not controlling in the experiment.

Another important parameter in analyzing our experiment is the overall average response. This quantity is obtained by summing all individual trial responses and dividing by the total number of trials. The calculation of the overall average experiment response is shown in the following equation:

$$\text{Average Response} = \overline{T} = \frac{\sum_{i=1}^{n} R_i}{n} = \frac{R_1 + R_2 + R_3 + R_4}{n}$$

$$\overline{T} = \frac{1.505 + 1.515 + 1.520 + 1.525}{4} \tag{10-5}$$

$$\overline{T} = 1.51625 \approx 1.5163$$

Response Graph

Now that we have calculated the factor level averages and the overall average response for the experiment, we are ready to graph the results. To construct our graph, we will place the control factor levels along the horizontal axis, and the quality characteristic response values along the vertical axis. We shall also plot the overall experiment average as a horizontal line on the graph (Figure 10-1).

The response graph provides a visual summary of the control factor effects. As we shall discover in later chapters, response graphs provide a great deal of information about individual control factors, and their relative importance in controlling a quality characteristic response. For now, it is sufficient to view the graph as a tool for summarizing results of the experimentation.

The graph plots of control factors A and B are called the plots of the *main effects* in our experimental array. A main effect is the amount of change in the quality characteristic response, which results directly from altering the control factor levels.

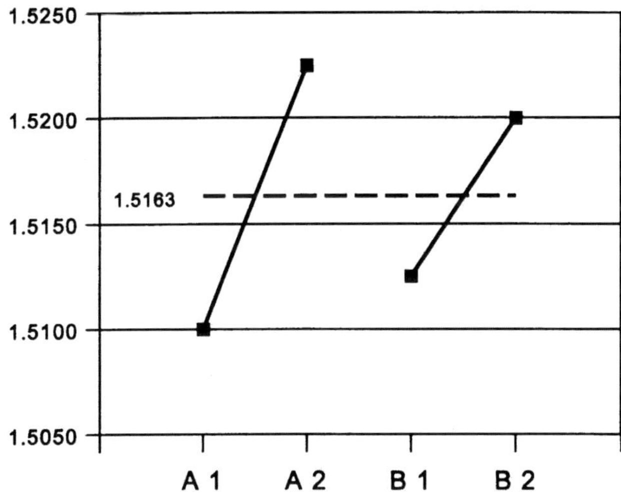

Figure 10-1: Response Graph

Larger Experimental Arrays

The array we created for two control factors is fine, but what if we have a desire to experiment with additional control factors? Can we create a larger experimental array in the same fashion? The answer is yes, and we will examine the creation of an array for three control factors.

Assume that we have identified a quality characteristic response, brainstormed and selected control factors and levels, and now have a need to experiment with three control factors at two levels. We shall use the 1, 2 notation as shown below.

Factor	Levels	
A	1	2
B	1	2
C	1	2

As we did earlier, we begin by creating a trial with all control factors at level 1, and adding a column to record the measured response of the quality characteristic we desire to study. We obtain:

Trial	A	B	C	Response
1	1	1	1	R_1

Concentrate initially on the columns for control factors B and C. We know from our earlier exercise that we must provide both levels of C to enable analysis of factor B at level 1. It is essential that the array be balanced. We therefore add another trial to the array with C at level 2:

Trial	A	B	C	Response
1	1	1	1	R_1
2	1	1	2	R_2

We have just "balanced" C for factor B at level 1. We must also provide both levels of factor C when factor B is at level 2. To accomplish this, we add two more trials to the array, both of which have control factor B at level 2. The new trials, 3 and 4, are laid out the same as 1 and 2, except we've changed the level of factor B. With this addition, the array now appears as illustrated on the following page.

Trial	A	B	C	Response
1	1	1	1	R_1
2	1	1	2	R_2
3	1	2	1	R_3
4	1	2	2	R_4

If we disregard the column for control factor A for a moment, we can see that the remaining portion of the array is exactly the same as the array we created earlier. We know from our earlier discussions the factors B and C are balanced. We may calculate an average for B_1 and C will have been run at both levels. We may also calculate an average for B_2 and C will have been run at both levels. Likewise, we may calculate averages for factor C at levels 1 and 2, and for each level of C, control factor B will have been run at two levels.

Trial	A	B	C	Response
1	1	1	1	R_1
2	1	1	2	R_2
3	1	2	1	R_3
4	1	2	2	R_4

If we now focus our attention on columns A and B as shown in the following illustration, we can see that A has thus far has only been run at level 1. We can see that B has been run twice at level 1 and twice at level 2. The calculation of an average effect for A_1 would therefore not be distorted by any single level of factor B. In other words, B is balanced for A at level 1.

Trial	A	B	C	Response
1	1	1	1	R_1
2	1	1	2	R_2
3	1	2	1	R_3
4	1	2	2	R_4

Next, if we focus on control factors A and C as shown below, we find that factor C has been run twice at level 1 and twice at level 2. This would enable us to calculate the average level effect for A_1 without any single level of factor C distorting the result.

Trial	A	B	C	Response
1	1	1	1	R_1
2	1	1	2	R_2
3	1	2	1	R_3
4	1	2	2	R_4

Our array for three control factor main effects is almost complete. In its present form, any calculation of average level effects for B or C would be based upon responses potentially affected by A at level 1. To eliminate the influence of A_1, we must run additional trials with B and C at the levels prescribed above, but with factor A at level 2. This is simple, since the layout of the array thus far is balanced with respect to factors B and C. We need only to duplicate the design, but establish control factor A at level 2. This array is shown on the following page.

We will not tabulate all of the equations for calculating average level effects at this time. A visual examination of the array should suffice to indicate that it is balanced for all control factors. For every control factor at one specific level, we will find that the other two factors have been specified twice at level 1 and twice at level 2.

Trial	A	B	C	Response
1	1	1	1	R_1
2	1	1	2	R_2
3	1	2	1	R_3
4	1	2	2	R_4
5	2	1	1	R_5
6	2	1	2	R_6
7	2	2	1	R_7
8	2	2	2	R_8

References

1. American Supplier Institute *Taguchi Methods, Introduction to Quality Engineering Course Manual,* Version 2.1, American Supplier Institute, Dearborn, MI, 1991.

2. Barker, T. B. *Engineering Quality by Design—Interpreting the Taguchi Approach*, Marcel Dekker, New York, 1990.

3. Barker, T. B. *Quality by Experimental Design,* 2nd edition, Marcel Dekker, New York, 1994.

4. Berger, R.W (ed.) and Pyzdek, T. (ed.) *Quality Engineering Handbook*, ASQC Quality Press, Milwaukee, WI, and Marcel Dekker, New York, 1992.

5. Lochner, R.H. and Matar, J.E. *Designing for Quality—An Introduction to the Best of Taguchi and Western Methods of Statistical Experimental Design*, Quality Resources, New York and ASQC Quality Press, 1990.

6. Montgomery, D.C. and Myers, R.H. *Response Surface Methodology —Process and Product Optimization Using Designed Experiments*, John Wiley & Sons, New York, 1995.

7. Taguchi, G. *System of Experimental Design,* UNIPUB/Kraus International Publications, White Plains, NY, and American Supplier Institute, Dearborn, MI, 1987.

8. Taguchi, G. *Introduction to Quality Engineering*, Asian Productivity Organization, Tokyo, 1986.

11

Control Factor Interactions

In the construction of our first experimental array in Chapter 10, we assumed control factors A and B were independent of one another. We assumed neither factor had any negative or positive influence upon the other when they were alternating between levels 1 and 2. Although this assumption is generally true, it is not always true for all manufacturing applications. What if there is an interaction between these factors that affects our experimental response? How do we determine if it exists? If one does exist, how do we measure the effect it has on the quality characteristic response?

Our next step is to examine the construction of interaction columns for our experimental arrays. We shall then learn how to analyze the experimental responses to determine whether or not an interaction exists between two control factors. We shall also examine a method to quantify the magnitude of an interaction effect. Finally, later in this chapter, we shall examine interaction effects for more than two control factors. While these higher order interactions are rare for all but chemical processes, the construction of columns to accommodate them in our experimental arrays is an important step in understanding the structure of the arrays.

Two-Factor Interactions

For two control factors, A and B, the interaction effect will be denoted as A×B (read A cross B) or simply AB. The AB interaction is classified as a two-factor interaction. To illustrate the AB interaction, we will make use of the first experimental array we created in Chapter 10. To determine if the AB interaction exists, we must modify the array we have constructed for control factors A and B. It is not necessary to "run" any additional experimental trials—we merely analyze whether or not an interaction has influenced the experimental responses already obtained. The experimental array (with response values) from Chapter 10 is repeated below.

Trial	A	B	Response
1	1	1	1.505
2	1	2	1.515
3	2	1	1.520
4	2	2	1.525

Our task is to construct an additional column in our array, labeled AB, and determine the appropriate levels for the interaction column. As we shall see, these levels are predetermined by the levels of the control factors, A and B. We stated in an earlier chapter that levels could be denoted in a variety of ways—either as +1, -1 or as 1, 2. To construct our interaction column we will replace the 1, 2 notation we are currently using with +1, -1 notation common to Western experimental designs. This notation will simplify the construction of the interaction column. To accomplish this, simply replace the 1's in our array with +1's and replace the 2's with -1's. We also insert a column for the AB interaction immediately after the column for factor B. Our new array is illustrated on the following page.

Trial	A	B	AB	Response
1	+1	+1		1.505
2	+1	-1		1.515
3	-1	+1		1.520
4	-1	-1		1.525

To determine the level of each trial in the AB column, simply multiply the A and B levels in each row. For example, in Trial 1, A is at the +1 level and B is at the +1 level. A×B yields $(+1) \times (+1) = +1$. So the level of the AB interaction effect for Trial 1 is +1. If we continue multiplying the A and B levels for each row, we obtain the array illustrated below:

Trial	A	B	AB	Response
1	+1	+1	+1	1.505
2	+1	-1	-1	1.515
3	-1	+1	-1	1.520
4	-1	-1	+1	1.525

The AB interaction column is balanced with respect to both control factor columns. For example, if we are calculating the average effect of factor A at level 1, we find the AB interaction occurs at both levels. If we were to examine the entire array as illustrated in the previous chapter, we would find this is true for factor A, for factor B, and for interaction AB.

To determine the magnitude of interaction effects, we block out unwanted rows and columns, just as we did when calculating the average effects of A_1, A_2, B_1, and B_2. We determine the average effect of the AB interaction at the +1 level as follows:

Trial	A	B	AB	Response
1	+1	+1	+1	1.505
2	+1	-1	-1	1.515
3	-1	+1	-1	1.520
4	-1	-1	+1	1.525

We denote the AB interaction effect at the +1 level as $(AB)_{+1}$ and calculate an average response as shown in Equation 11-1.

$$(AB)_{+1} = \frac{1.505 + 1.525}{2}$$

$$(AB)_{+1} = 1.515$$

(11-1)

We determine the AB interaction effect at the -1 level as follows:

Trial	A	B	AB	Response
1	+1	+1	+1	1.505
2	+1	-1	-1	1.515
3	-1	+1	-1	1.520
4	-1	-1	+1	1.525

$$(AB)_{-1} = \frac{1.515 + 1.520}{2}$$

$$(AB)_{-1} = 1.5175$$

(11-2)

After calculating the average response for each level of the interaction, we may calculate the magnitude of the AB interaction effect by simply subtracting one result from the other. For uniformity, let's return to the 1, 2 notation we utilized in our original array. $(AB)_{-1}$ is denoted by $(AB)_2$ and $(AB)_{+1}$ is denoted by $(AB)_1$. The AB interaction may now be included in our response table and response graph as follows:

	Control Factor A	Control Factor B	Interaction AB
Level 2 Average	1.5225	1.5200	1.5175
Level 1 Average	1.5100	1.5125	1.5150
Average Difference	.0125	.0075	.0025

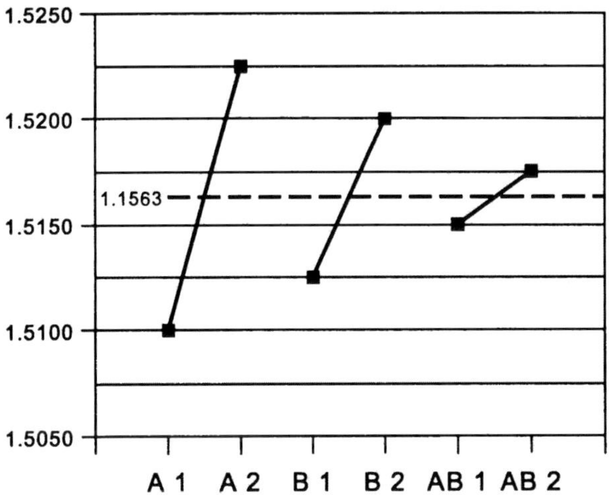

Figure 11-1: Response Graph with Interaction Effect

The inclusion of the interaction effect on our response graph illustrates the magnitude of the interaction effect is smaller than both main effects. It appears to be less significant than either control factor A or control factor B in changing the magnitude of the quality characteristic response we were studying, part length.

Interaction Graph

To fully investigate the significance of the AB interaction effect, an additional graph is needed. This "interaction graph" enables us to select the levels of A and B necessary to optimize our response when the interaction effect is significant. To construct the two-factor interaction graph, we:

1) determine the average response for each unique factor combination
2) use one set of control factor levels as labels on the horizontal axis and the response measurement as the vertical axis
3) plot the average response data as points on the graph
4) connect the points corresponding to each level of the second control factor and label the lines

For example, from our current experimental array (repeated on the next page) we obtain the following data:

$$A_1B_1 = 1.505$$

$$A_1B_2 = 1.515$$

$$A_2B_1 = 1.520$$

$$A_2B_2 = 1.525$$

Trial	A	B	AB	Response
1	1	1	1	1.505
2	1	2	2	1.515
3	2	1	2	1.520
4	2	2	1	1.525

We now construct a graph with factor A on the horizontal axis and the response variable on the vertical axis. We plot the four data points, connect the points corresponding to each level of control factor B and label the lines.

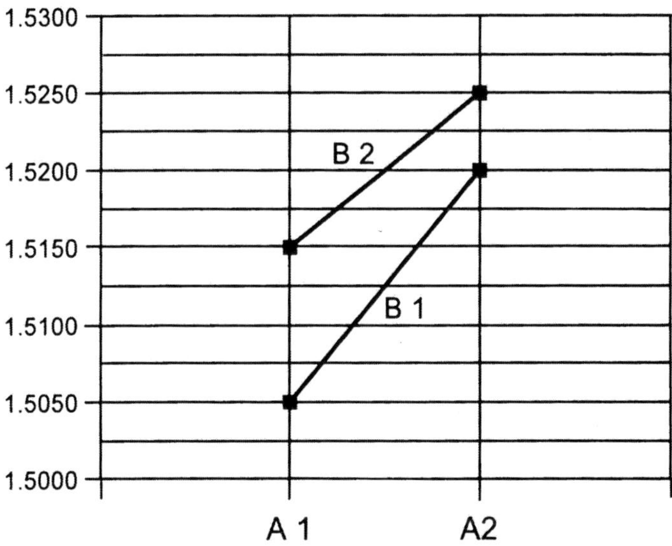

Figure 11-2: AB Interaction Graph

The slope of the B_1 and B_2 lines, with respect to one another, is fundamental in the analysis of the interaction graph. If the lines are parallel, no interaction exists for the factors A and B. Changing control factor A from level 1 to level 2 has the same effect on the output regardless of the level of control factor B. If the lines are not parallel, an interaction does exist for A and B. The strength of the interaction is indicated by the degree of non-perpendicularity, with the interaction between the factors stronger as the lines approach perpendicular.

The average responses for AB_1 and AB_2 were calculated from the response column of our experimental array. We then added a column to the response table and calculated the magnitude of the interaction (or average difference) to be .0025 inches. The interaction effect can be represented graphically as shown in Figure 11-3.

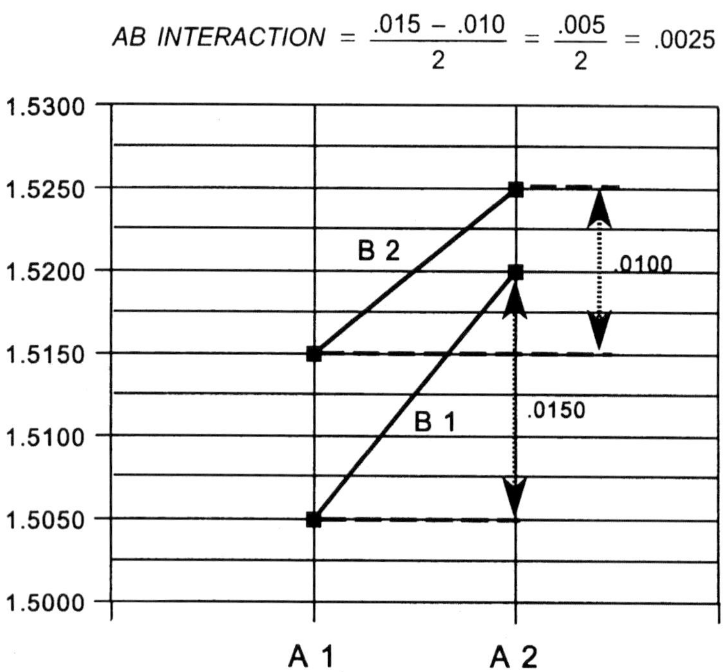

Figure 11-3: AB Interaction Magnitude

A few general comments about interaction effects are appropriate:

1) Interactions are usually not significant unless at least one of the main effects is also significant.
2) Use the response graph to determine whether or not an interaction graph is necessary.
3) If a significant interaction is present, use the interaction graph to establish the preferred levels for the main effects.

In our example, the interaction effect was .0025, while the main effects were .125 and .075 for factors A and B, respectively. Relative to the main effects, we would consider the interaction effect insignificant.

Higher Order Interactions

In the previous chapter we created the following array for an experiment with three control factors, A, B, C, each at two levels.

Trial	A	B	C	Response
1	1	1	1	R_1
2	1	1	2	R_2
3	1	2	1	R_3
4	1	2	2	R_4
5	2	1	1	R_5
6	2	1	2	R_6
7	2	2	1	R_7
8	2	2	2	R_8

Careful examination of this array will reveal *four* possible interactions, even though the array contains only three control factors. If we begin with control factor A, and search for all possible two-factor interactions to the right, we immediately see that an AB interaction is possible: $A \times B \rightarrow AB$. We also note that an AC interaction is possible: $A \times C \rightarrow AC$. Since we have exhausted all single factors to the right of A, now begin with B and search to the right for interactions. The only possible interaction in this iteration is: $B \times C \rightarrow BC$. There are no factors to the right of C, therefore no additional two-factor interactions are possible from this array. Thus far, we have identified the following:

$$A \times B \rightarrow AB$$
$$A \times C \rightarrow AC$$
$$B \times C \rightarrow BC$$

One additional interaction is possible from this array—this is the three-factor interaction—$A \times B \times C \rightarrow ABC$. The four possible interactions which may be analyzed with this array are: AB, AC, BC, and ABC. To construct the columns for the interaction effects, we once again switch from the 1, 2 notation used in the above table to the -1, +1 notation.

Trial	A	B	C	Response
1	+1	+1	+1	R_1
2	+1	+1	-1	R_2
3	+1	-1	+1	R_3
4	+1	-1	-1	R_4
5	-1	+1	+1	R_5
6	-1	+1	-1	R_6
7	-1	-1	+1	R_7
8	-1	-1	-1	R_8

Since we have determined four interactions were possible, we add four columns to our experimental array and label each of them with one of the possible the interactions.

Trial	A	B	C	AB	AC	BC	ABC	Response
1	+1	+1	+1					R_1
2	+1	+1	-1					R_2
3	+1	-1	+1					R_3
4	+1	-1	-1					R_4
5	-1	+1	+1					R_5
6	-1	+1	-1					R_6
7	-1	-1	+1					R_7
8	-1	-1	-1					R_8

As we stated before: *We do not run interactions*. Their levels are predetermined by the levels of control factors A, B, and C in the array shown above. For example, to find the levels of the AB interaction for each trial of the experiment, we simply multiply the corresponding levels of factors A and B.

The following pages illustrate the calculation of interaction effects for this experimental array. The technique is valid for any of the two-level arrays examined in this text. We begin by showing the calculation of both the AB and AC interaction levels on the following page. Subsequent pages show the calculation for the BC and the ABC interaction effects, and the completed experimental array with all of the interaction levels.

Calculation of AB Interaction

Trial	A	B	C	AB	AC	BC	ABC	Response
1	+1	+1	+1	+1				R_1
2	+1	+1	-1	+1				R_2
3	+1	-1	+1	-1				R_3
4	+1	-1	-1	-1				R_4
5	-1	+1	+1	-1				R_5
6	-1	+1	-1	-1				R_6
7	-1	-1	+1	+1				R_7
8	-1	-1	-1	+1				R_8

The AC interaction is determined in the same fashion. Consider only columns A and C, then multiply their respective levels for each trial to determine the level of the AC interaction effect.

Calculation of AC Interaction

Trial	A	B	C	AB	AC	BC	ABC	Response
1	+1	+1	+1	+1	+1			R_1
2	+1	+1	-1	+1	-1			R_2
3	+1	-1	+1	-1	+1			R_3
4	+1	-1	-1	-1	-1			R_4
5	-1	+1	+1	-1	-1			R_5
6	-1	+1	-1	-1	+1			R_6
7	-1	-1	+1	+1	-1			R_7
8	-1	-1	-1	+1	+1			R_8

The BC interaction is found by multiplying the levels of factors B and C in each trial, as shown below:

Calculation of BC Interaction

Trial	A	B	C	AB	AC	BC	ABC	Response
1	+1	+1	+1	+1	+1	+1		R_1
2	+1	+1	-1	+1	-1	-1		R_2
3	+1	-1	+1	-1	+1	-1		R_3
4	+1	-1	-1	-1	-1	+1		R_4
5	-1	+1	+1	-1	-1	+1		R_5
6	-1	+1	-1	-1	+1	-1		R_6
7	-1	-1	+1	+1	-1	-1		R_7
8	-1	-1	-1	+1	+1	+1		R_8

The ABC interaction if found by multiplying the levels of factors A, B, and C in each trial:

Calculation of ABC Interaction

Trial	A	B	C	AB	AC	BC	ABC	Response
1	+1	+1	+1	+1	+1	+1	+1	R_1
2	+1	+1	-1	+1	-1	-1	-1	R_2
3	+1	-1	+1	-1	+1	-1	-1	R_3
4	+1	-1	-1	-1	-1	+1	+1	R_4
5	-1	+1	+1	-1	-1	+1	-1	R_5
6	-1	+1	-1	-1	+1	-1	+1	R_6
7	-1	-1	+1	+1	-1	-1	+1	R_7
8	-1	-1	-1	+1	+1	+1	-1	R_8

Our completed array, with interaction columns, appears as follows:

Trial	A	B	C	AB	AC	BC	ABC	Response
1	+1	+1	+1	+1	+1	+1	+1	R_1
2	+1	+1	-1	+1	-1	-1	-1	R_2
3	+1	-1	+1	-1	+1	-1	-1	R_3
4	+1	-1	-1	-1	-1	+1	+1	R_4
5	-1	+1	+1	-1	-1	+1	-1	R_5
6	-1	+1	-1	-1	+1	-1	+1	R_6
7	-1	-1	+1	+1	-1	-1	+1	R_7
8	-1	-1	-1	+1	+1	+1	-1	R_8

In the beginning of this chapter we created an AB interaction column for the array we had created in Chapter 10. Using the response column from our sample array, we then calculated the average effect of $(AB)_{-1}$ and $(AB)_{+1}$. These interaction effects were then included in a response table and in a response graph. If we were using the above array for experimentation, the same type of analysis is required for all four of the interactions we have just studied and for the three control factors, A, B, and C. This analysis will be detailed in subsequent chapters.

While two-factor interactions are sometimes found to be significant in industrial applications, most three-factor and higher order interactions are seldom significant unless we are dealing with a chemical reaction. Also, as stated earlier, we will not normally find interactions are significant unless one or more of the control factors are also significant.

References

1. American Supplier Institute *Taguchi Methods, Introduction to Quality Engineering Course Manual,* Version 2.1, American Supplier Institute, Dearborn, MI, 1991.

2. Barker, T. B. *Engineering Quality by Design—Interpreting the Taguchi Approach*, Marcel Dekker, New York, 1990.

3. Barker, T. B. *Quality by Experimental Design,* 2nd edition, Marcel Dekker, New York, 1994.

4. Berger, R.W (ed.) and Pyzdek, T. (ed.) *Quality Engineering Handbook*, ASQC Quality Press, Milwaukee, WI, and Marcel Dekker, New York, 1992.

5. Lochner, R.H. and Matar, J.E. *Designing for Quality—An Introduction to the Best of Taguchi and Western Methods of Statistical Experimental Design*, Quality Resources, New York and ASQC Quality Press, 1990.

6. Montgomery, D.C. and Myers, R.H. *Response Surface Methodology —Process and Product Optimization Using Designed Experiments,* John Wiley & Sons, New York, 1995.

7. Taguchi, G. *System of Experimental Design,* UNIPUB/Kraus International Publications, White Plains, NY, and American Supplier Institute, Dearborn, MI, 1987.

8. Taguchi, G. *Introduction to Quality Engineering*, Asian Productivity Organization, Tokyo, 1986.

12

Orthogonal Arrays

An orthogonal array is one which allows the analysis of factor effects in a mathematically independent fashion. As a result, the analysis of a factor in any column will be independent of the factors in the other columns. In practical terms, it is a balanced array. In practice, we will not normally be concerned with testing for orthogonality. Typically, we will determine the number of control factors and/or interactions we want to evaluate and select an experimental array from a reference source (arrays are available in the Appendix). We then design our manufacturing trials according to the constraints of the selected array. Still, it is important to understand the concept of orthogonality and have some simple criteria for testing if we are uncertain whether or not an array is balanced.

We constructed our own experimental arrays in previous chapters and took much care to ensure that they were balanced. To determine if a two-level array is balanced, use the -1, +1 notation common to Western arrays, multiply the row values of any two columns together and compute the sum of the products. The two columns are orthogonal to one another if the sum of their row products is equal to zero. To illustrate, let's study the experimental array for control factors A and B that we created in Chapter 10.

Trial	A	B	Response
1	1	1	R_1
2	1	2	R_2
3	2	1	R_3
4	2	2	R_4

If we examine this array in -1, +1 notation and replace the response column with a product column we obtain the following:

Trial	A	B	Product
1	+1	+1	
2	+1	-1	
3	-1	+1	
4	-1	-1	

Now, multiply the level of A times the level of B in each row, record the product and calculate the sum of the products.

Trial	A	B	Product
1	+1	+1	+1
2	+1	-1	-1
3	-1	+1	-1
4	-1	-1	+1
		Sum	0

When we add the products of the A and B columns, we obtain a sum of zero. This meets the definition for orthogonality, therefore, we may state columns A and B are orthogonal (or balanced) with respect to each another. Notice the product column is equal to A×B—this is the same value we calculated to assess the level of the AB interaction effect.

We also created an experimental array for three control factors at two levels in an earlier chapter (we shall refer to this as the ABC array). We then switched for 1, 2 notation to +1, -1 notation and constructed interaction columns for AB, AC, BC, and ABC. This ABC array is repeated below:

Trial	A	B	C	AB	AC	BC	ABC	Response
1	+1	+1	+1	+1	+1	+1	+1	R_1
2	+1	+1	-1	+1	-1	-1	-1	R_2
3	+1	-1	+1	-1	+1	-1	-1	R_3
4	+1	-1	-1	-1	-1	+1	+1	R_4
5	-1	+1	+1	-1	-1	+1	-1	R_5
6	-1	+1	-1	-1	+1	-1	+1	R_6
7	-1	-1	+1	+1	-1	-1	+1	R_7
8	-1	-1	-1	+1	+1	+1	-1	R_8

Since we constructed the AB interaction column by multiplying the level of A and the level of B in each row (A×B), there is no need to repeat the exercise to determine if the columns are orthogonal. If the sum of this column is equal to zero, we may state that columns A and B are orthogonal to one another. The same test also applies to the AC, BC, and the ABC columns. We add the column values to test for orthogonality on the following page:

Trial	A	B	C	AB	AC	BC	ABC
1	+1	+1	+1	+1	+1	+1	+1
2	+1	+1	-1	+1	-1	-1	-1
3	+1	-1	+1	-1	+1	-1	-1
4	+1	-1	-1	-1	-1	+1	+1
5	-1	+1	+1	-1	-1	+1	-1
6	-1	+1	-1	-1	+1	-1	+1
7	-1	-1	+1	+1	-1	-1	+1
8	-1	-1	-1	+1	+1	+1	-1
Sum of Products				0	0	0	0

Since the sums of the products are equal to zero, we may say the columns are orthogonal to one another. Once again, this concept is important because it allows us to calculate the average effects of the column factors individually, without concern that the results of our calculations have been effected by the other factors. In other words, for any calculation of an average level effect for one factor, all other factors have been run an equal number of times at both the +1 and the -1 level.

When we created the ABC array we did not attempt to calculate average level effects for the control factors or the interactions. It is a very useful exercise, since the ABC array is one of the most common arrays used in manufacturing process optimization efforts. We will examine these average level effects on the following few pages. We will not use numeric values, but we will add the response column to the array, and formulate equations in terms of the response value, R_n.

Trial	A	B	C	AB	AC	BC	ABC	Response
1	+1	+1	+1	+1	+1	+1	+1	R_1
2	+1	+1	-1	+1	-1	-1	-1	R_2
3	+1	-1	+1	-1	+1	-1	-1	R_3
4	+1	-1	-1	-1	-1	+1	+1	R_4
5	-1	+1	+1	-1	-1	+1	-1	R_5
6	-1	+1	-1	-1	+1	-1	+1	R_6
7	-1	-1	+1	+1	-1	-1	+1	R_7
8	-1	-1	-1	+1	+1	+1	-1	R_8

$$A_{+1} = \frac{R_1 + R_2 + R_3 + R_4}{4} \qquad (12\text{-}1)$$

Trial	A	B	C	AB	AC	BC	ABC	Response
1	+1	+1	+1	+1	+1	+1	+1	R_1
2	+1	+1	-1	+1	-1	-1	-1	R_2
3	+1	-1	+1	-1	+1	-1	-1	R_3
4	+1	-1	-1	-1	-1	+1	+1	R_4
5	-1	+1	+1	-1	-1	+1	-1	R_5
6	-1	+1	-1	-1	+1	-1	+1	R_6
7	-1	-1	+1	+1	-1	-1	+1	R_7
8	-1	-1	-1	+1	+1	+1	-1	R_8

$$A_{-1} = \frac{R_5 + R_6 + R_7 + R_8}{4} \qquad (12\text{-}2)$$

Trial	A	B	C	AB	AC	BC	ABC	Response
1	+1	+1	+1	+1	+1	+1	+1	R_1
2	+1	+1	-1	+1	-1	-1	-1	R_2
3	+1	-1	+1	-1	+1	-1	-1	R_3
4	+1	-1	-1	-1	-1	+1	+1	R_4
5	-1	+1	+1	-1	-1	+1	-1	R_5
6	-1	+1	-1	-1	+1	-1	+1	R_6
7	-1	-1	+1	+1	-1	-1	+1	R_7
8	-1	-1	-1	+1	+1	+1	-1	R_8

$$B_{+1} = \frac{R_1 + R_2 + R_5 + R_6}{4} \qquad (12\text{-}3)$$

Trial	A	B	C	AB	AC	BC	ABC	Response
1	+1	+1	+1	+1	+1	+1	+1	R_1
2	+1	+1	-1	+1	-1	-1	-1	R_2
3	+1	-1	+1	-1	+1	-1	-1	R_3
4	+1	-1	-1	-1	-1	+1	+1	R_4
5	-1	+1	+1	-1	-1	+1	-1	R_5
6	-1	+1	-1	-1	+1	-1	+1	R_6
7	-1	-1	+1	+1	-1	-1	+1	R_7
8	-1	-1	-1	+1	+1	+1	-1	R_8

$$B_{-1} = \frac{R_3 + R_4 + R_7 + R_8}{4} \qquad (12\text{-}4)$$

Trial	A	B	C	AB	AC	BC	ABC	Response
1	+1	+1	+1	+1	+1	+1	+1	R_1
2	+1	+1	-1	+1	-1	-1	-1	R_2
3	+1	-1	+1	-1	+1	-1	-1	R_3
4	+1	-1	-1	-1	-1	+1	+1	R_4
5	-1	+1	+1	-1	-1	+1	-1	R_5
6	-1	+1	-1	-1	+1	-1	+1	R_6
7	-1	-1	+1	+1	-1	-1	+1	R_7
8	-1	-1	-1	+1	+1	+1	-1	R_8

$$C_{+1} = \frac{R_1 + R_3 + R_5 + R_7}{4} \tag{12-5}$$

Trial	A	B	C	AB	AC	BC	ABC	Response
1	+1	+1	+1	+1	+1	+1	+1	R_1
2	+1	+1	-1	+1	-1	-1	-1	R_2
3	+1	-1	+1	-1	+1	-1	-1	R_3
4	+1	-1	-1	-1	-1	+1	+1	R_4
5	-1	+1	+1	-1	-1	+1	-1	R_5
6	-1	+1	-1	-1	+1	-1	+1	R_6
7	-1	-1	+1	+1	-1	-1	+1	R_7
8	-1	-1	-1	+1	+1	+1	-1	R_8

$$C_{-1} = \frac{R_2 + R_4 + R_6 + R_8}{4} \tag{12-6}$$

Trial	A	B	C	AB	AC	BC	ABC	Response
1	+1	+1	+1	+1	+1	+1	+1	R_1
2	+1	+1	-1	+1	-1	-1	-1	R_2
3	+1	-1	+1	-1	+1	-1	-1	R_3
4	+1	-1	-1	-1	-1	+1	+1	R_4
5	-1	+1	+1	-1	-1	+1	-1	R_5
6	-1	+1	-1	-1	+1	-1	+1	R_6
7	-1	-1	+1	+1	-1	-1	+1	R_7
8	-1	-1	-1	+1	+1	+1	-1	R_8

$$(AB)_{+1} = \frac{R_1 + R_2 + R_7 + R_8}{4} \tag{12-7}$$

Trial	A	B	C	AB	AC	BC	ABC	Response
1	+1	+1	+1	+1	+1	+1	+1	R_1
2	+1	+1	-1	+1	-1	-1	-1	R_2
3	+1	-1	+1	-1	+1	-1	-1	R_3
4	+1	-1	-1	-1	-1	+1	+1	R_4
5	-1	+1	+1	-1	-1	+1	-1	R_5
6	-1	+1	-1	-1	+1	-1	+1	R_6
7	-1	-1	+1	+1	-1	-1	+1	R_7
8	-1	-1	-1	+1	+1	+1	-1	R_8

$$(AB)_{-1} = \frac{R_3 + R_4 + R_5 + R_6}{4} \tag{12-8}$$

Trial	A	B	C	AB	AC	BC	ABC	Response
1	+1	+1	+1	+1	+1	+1	+1	R_1
2	+1	+1	-1	+1	-1	-1	-1	R_2
3	+1	-1	+1	-1	+1	-1	-1	R_3
4	+1	-1	-1	-1	-1	+1	+1	R_4
5	-1	+1	+1	-1	-1	+1	-1	R_5
6	-1	+1	-1	-1	+1	-1	+1	R_6
7	-1	-1	+1	+1	-1	-1	+1	R_7
8	-1	-1	-1	+1	+1	+1	-1	R_8

$$(AC)_{+1} = \frac{R_1 + R_3 + R_6 + R_8}{4} \qquad (12\text{-}9)$$

Trial	A	B	C	AB	AC	BC	ABC	Response
1	+1	+1	+1	+1	+1	+1	+1	R_1
2	+1	+1	-1	+1	-1	-1	-1	R_2
3	+1	-1	+1	-1	+1	-1	-1	R_3
4	+1	-1	-1	-1	-1	+1	+1	R_4
5	-1	+1	+1	-1	-1	+1	-1	R_5
6	-1	+1	-1	-1	+1	-1	+1	R_6
7	-1	-1	+1	+1	-1	-1	+1	R_7
8	-1	-1	-1	+1	+1	+1	-1	R_8

$$(AC)_{-1} = \frac{R_2 + R_4 + R_5 + R_7}{4} \qquad (12\text{-}10)$$

Trial	A	B	C	AB	AC	BC	ABC	Response
1	+1	+1	+1	+1	+1	+1	+1	R_1
2	+1	+1	-1	+1	-1	-1	-1	R_2
3	+1	-1	+1	-1	+1	-1	-1	R_3
4	+1	-1	-1	-1	-1	+1	+1	R_4
5	-1	+1	+1	-1	-1	+1	-1	R_5
6	-1	+1	-1	-1	+1	-1	+1	R_6
7	-1	-1	+1	+1	-1	-1	+1	R_7
8	-1	-1	-1	+1	+1	+1	-1	R_8

$$(BC)_{+1} = \frac{R_1 + R_4 + R_5 + R_8}{4} \qquad (12\text{-}11)$$

Trial	A	B	C	AB	AC	BC	ABC	Response
1	+1	+1	+1	+1	+1	+1	+1	R_1
2	+1	+1	-1	+1	-1	-1	-1	R_2
3	+1	-1	+1	-1	+1	-1	-1	R_3
4	+1	-1	-1	-1	-1	+1	+1	R_4
5	-1	+1	+1	-1	-1	+1	-1	R_5
6	-1	+1	-1	-1	+1	-1	+1	R_6
7	-1	-1	+1	+1	-1	-1	+1	R_7
8	-1	-1	-1	+1	+1	+1	-1	R_8

$$(BC)_{-1} = \frac{R_2 + R_3 + R_6 + R_7}{4} \qquad (12\text{-}12)$$

Trial	A	B	C	AB	AC	BC	ABC	Response
1	+1	+1	+1	+1	+1	+1	+1	R_1
2	+1	+1	-1	+1	-1	-1	-1	R_2
3	+1	-1	+1	-1	+1	-1	-1	R_3
4	+1	-1	-1	-1	-1	+1	+1	R_4
5	-1	+1	+1	-1	-1	+1	-1	R_5
6	-1	+1	-1	-1	+1	-1	+1	R_6
7	-1	-1	+1	+1	-1	-1	+1	R_7
8	-1	-1	-1	+1	+1	+1	-1	R_8

$$(ABC)_{+1} = \frac{R_1 + R_4 + R_6 + R_7}{4} \qquad (12\text{-}13)$$

Trial	A	B	C	AB	AC	BC	ABC	Response
1	+1	+1	+1	+1	+1	+1	+1	R_1
2	+1	+1	-1	+1	-1	-1	-1	R_2
3	+1	-1	+1	-1	+1	-1	-1	R_3
4	+1	-1	-1	-1	-1	+1	+1	R_4
5	-1	+1	+1	-1	-1	+1	-1	R_5
6	-1	+1	-1	-1	+1	-1	+1	R_6
7	-1	-1	+1	+1	-1	-1	+1	R_7
8	-1	-1	-1	+1	+1	+1	-1	R_8

$$(ABC)_{-1} = \frac{R_2 + R_3 + R_5 + R_8}{4} \qquad (12\text{-}14)$$

$$A_{+1} = (R_1 + R_2 + R_3 + R_4) / 4$$

$$A_{-1} = (R_5 + R_6 + R_7 + R_8) / 4$$

$$B_{+1} = (R_1 + R_2 + R_5 + R_6) / 4$$

$$B_{-1} = (R_3 + R_4 + R_7 + R_8) / 4$$

$$C_{+1} = (R_1 + R_3 + R_5 + R_7) / 4$$

$$C_{-1} = (R_2 + R_4 + R_6 + R_8) / 4$$

$$(AB)_{+1} = (R_3 + R_4 + R_5 + R_6) / 4$$

$$(AB)_{-1} = (R_1 + R_2 + R_7 + R_8) / 4$$

$$(AC)_{+1} = (R_2 + R_4 + R_5 + R_7) / 4 \qquad (12\text{-}15)$$

$$(AC)_{-1} = (R_1 + R_3 + R_6 + R_8) / 4$$

$$(BC)_{+1} = (R_2 + R_3 + R_6 + R_7) / 4$$

$$(BC)_{-1} = (R_1 + R_4 + R_5 + R_8) / 4$$

$$(ABC)_{+1} = (R_1 + R_4 + R_6 + R_7) / 4$$

$$(ABC)_{-1} = (R_2 + R_3 + R_5 + R_8) / 4$$

The equations for all factor and interaction average effects have been summarized above. Notice that each average level effect has a unique equation. This is a result of the orthogonal arrangement of this array. *These equations are valid only for an array with levels arranged as shown in our example.* They have been provided to aid in the visualization of the average level effect calculation technique.

We have now completed two arrays with interaction columns. The first was for two control factors at two levels, A and B with the AB interaction column. The second was for three control factors at two levels, A B and C with the AB, AC, BC, and ABC interaction columns. For industrial experimentation, these two experimental arrays will suffice for the vast majority of investigations. The ABC array, as we shall explain later, is most often the preferred method.

Experimental Array Notation

We have been labeling the rows in our arrays as trials. This is acceptable, and even preferred in an industrial setting for communication with operating personnel. But we must also be aware that many reference materials and texts will label the rows in terms of treatment combinations. Consider our AB array which included the AB interaction.

Trial	A	B	AB
1	+1	+1	+1
2	+1	-1	-1
3	-1	+1	-1
4	-1	-1	+1

In reference materials for Western experimentation, this array would appear as follows:

tc	A	B	AB
(1)	-1	-1	+1
a	+1	-1	-1
b	-1	+1	-1
ab	+1	+1	+1

The column heading has been changed from "Trial" to "tc," which stands for treatment combination, and refers to the combination of control factor levels to which the process is going to be subjected. The (1) indicates the base run, where all control factors are set at the -1 level. The "a" indicates the level of control factor A is set at +1 in this row. The "b" indicates control factor b is set at +1 in this row, and the "ab" indicates both control factors A and B are set at +1 in this row. (If we memorize the order of the "tc" column, the array can be constructed from memory.)

Notice that the order of the trials is different in the two arrays. The first trial in the top array (trial 1) is the same as the last trial in the bottom array (treatment combination ab). The last trial in the top array (trial 4) is the same as the first trial in the bottom array (the base run). The product of the A and B columns remains the same for both arrays, and the sum of the products is equal to zero in both cases. This indicates both arrays are orthogonal. The differences are in order and notation only. In fact, different order and notation is possible from different reference works. This probably generates much of the confusion many experience with experimental design techniques. The Taguchi L_4 design[4], for instance, appears as follows:

No.	1	2	3
1	1	1	1
2	1	2	2
3	2	1	2
4	2	2	1

If we replace the 1's in this array with +1's, and replace the 2's with -1's, we obtain the array illustrated on the following page. This array is the same as the one we created. The difference in notation, depending upon the source, is the principle reason we have spent so much time switching from -1's and +1's to 1's and 2's. We must be comfortable with the differing notations when we are reviewing arrays or experiments from different sources. The notation does not matter—it is only important that

---

[4] ©Copyright, American Supplier Institute, Inc., Allen Park, Michigan (U.S.A.) "Reproduced by permission under License No. 970101."

the designs we are reviewing are orthogonal (other designs are possible, but they are beyond the scope of this text).

No.	1	2	3
1	+1	+1	+1
2	+1	-1	-1
3	-1	+1	-1
4	-1	-1	+1

With respect to notation, you will typically design an experiment and label the control factor levels in accordance with an experimental array you have selected from a reference source. Once you have selected your array, stick with the published notation. Use whichever notation you feel most comfortable with. A few experimental arrays with appropriate notation are listed in the Appendix. These are by no means all inclusive, however, they will suffice for most experiments. Many other possibilities will be found in the listed reference materials at the end of each chapter.

References

1. American Supplier Institute *Taguchi Methods, Introduction to Quality Engineering Course Manual,* Version 2.1, American Supplier Institute, Dearborn, MI, 1991.

2. Barker, T. B. *Engineering Quality by Design—Interpreting the Taguchi Approach*, Marcel Dekker, New York, 1990.

3. Barker, T. B. *Quality by Experimental Design,* 2nd edition, Marcel Dekker, New York, 1994.

4. Berger, R.W (ed.) and Pyzdek, T. (ed.) *Quality Engineering Handbook*, ASQC Quality Press, Milwaukee, WI, and Marcel Dekker, New York, 1992.

5. Lochner, R.H. and Matar, J.E. *Designing for Quality—An Introduction to the Best of Taguchi and Western Methods of Statistical Experimental Design*, Quality Resources, New York and ASQC Quality Press, 1990.

6. Montgomery, D.C. and Myers, R.H. *Response Surface Methodology—Process and Product Optimization Using Designed Experiments*, John Wiley & Sons, New York, 1995.

7. Taguchi, G. *System of Experimental Design*, UNIPUB/Kraus International Publications, White Plains, NY, and American Supplier Institute, Dearborn, MI, 1987.

8. Taguchi, G. *Introduction to Quality Engineering*, Asian Productivity Organization, Tokyo, 1986.

13

Full Factorial Arrays

In Chapter 12 we discussed orthogonal arrays as "balanced" arrays. We introduced a test to determine if columns in an array are orthogonal to one another, and then determined that our sample AB and our sample ABC arrays were orthogonal. We also examined notation, and compared the classical or Western experimental design notation to the notation in Taguchi style arrays. All of this is important to our understanding of experimental arrays. However, our knowledge would be seriously lacking unless we discussed a few more fundamentals prior to actually designing and running experiments.

Two key concepts of designed experimentation we shall explore in this and the next chapter are full factorial arrays and fractional factorial arrays. As we shall see, even though both can utilize the same arrays, they represent two different and important methods of experimentation. We begin in this chapter by examining the full factorial array. Since this knowledge would provide us with the ability to actually initiate the design of experiments for process improvement or optimization efforts, we will then discuss a few additional concepts such as replication, randomization, and blocking.

Full Factorial Arrays

For our purposes, the full factorial array is defined as follows:

*A full factorial array is one which examines the average
level effects of all control factors and their interactions.*

This method of experimentation requires the use of orthogonal
arrays. The number of treatment combinations needed for a full factorial,
2-level array is shown in Equation 13-1 below.

$$treatment\ combinations\ =\ tc\ =\ 2^k$$

$$where:\ k\ =\ number\ of\ factors \tag{13-1}$$

As we saw when reviewing notation for the Western array in the
previous unit, treatment combinations are the number of trials needed for
the experiment to evaluate all the combinations of the factors being
examined. For an experiment with two control factors, the number of
treatment combinations or trials is 2^2 or 4. For three control factors the
number of tc's is 2^3 or 8, etc.

When we added an interaction column to our AB array in an earlier
chapter, the array became a full factorial design. Since there were only 2
control factors in our array, only one first order interaction was possible,
AxB or AB.

Trial	A	B	AB
1	+1	+1	+1
2	+1	-1	-1
3	-1	+1	-1
4	-1	-1	+1

In the ABC array we created, four interaction columns were needed for the array to be considered a full factorial design.

Trial	A	B	C	AB	AC	BC	ABC
1	+1	+1	+1	+1	+1	+1	+1
2	+1	+1	-1	+1	-1	-1	-1
3	+1	-1	+1	-1	+1	-1	-1
4	+1	-1	-1	-1	-1	+1	+1
5	-1	+1	+1	-1	-1	+1	-1
6	-1	+1	-1	-1	+1	-1	+1
7	-1	-1	+1	+1	-1	-1	+1
8	-1	-1	-1	+1	+1	+1	-1

With this type of design we run each control factor at the selected levels and calculate the response for all control factors, and all interactions. With respect to the two levels in the design, we run all possible combinations. This provides a response for all possible control factor settings. (If any of the control factors are continuous, there are an infinite number of possible levels at which the process may be run. The phrase "all possible control factor settings" as used here refers to the possible combinations of control factors at the discrete levels selected.)

The full factorial array is an excellent tool for increasing knowledge of the process under study. Within the range of control factor levels selected, it provides insight into which factors control the process, and the magnitude and direction of their effect. It also allows us to determine whether or not any factors interact in a positive or negative fashion with any of the other factors in the array. At times, this information can be critical to manufacturing optimization efforts. Utilization of orthogonal arrays in this fashion is ideal for R&D efforts, and for new product and process development efforts. It can be an extremely powerful tool in developing specifications for processing and purchasing. While it is critical that we know which factors actually control our processes, it is also important to know which factors have no discernable effects on our output characteristics. By loosening or eliminating specifications on such factors we avoid the expense associated with their control.

The principal drawback of the full factorial array, at least in a manufacturing environment, is the time and expense involved in running this type of experiment. It is not at all uncommon when brainstorming a complex process to identify many more potential control factors than the three which may be tested in an full factorial array with eight treatment combinations. While larger arrays are certainly available, for a two-level experiment the number of treatment combinations grows according to the 2^k (or L^n) relationship presented earlier.

Number of Control Factors	Treatment Combinations Required
2	4
3	8
4	16
5	32
6	64
7	128
8	256

The problem in a typical facility is that once a problem arises, it adversely effects production, quality, and shipments. There is rarely enough time to embark on prolonged testing to develop a more complete understanding of the process. Also, as we shall explore in the following section, it is unwise to perform each experimental trial only one time.

Replication

Until now, we have examined arrays with a single response column. This allowed us to concentrate on average effects for control factor and interaction levels. Industrial experiments should never be accomplished with one response value for each trial of the experiment. Variation in inputs (even in controlled inputs), the presence of noise factors, and the possibility of control factors undocumented or unknown to the experimenters all combine to produce variation in the characteristic responses being measured.

To estimate the amount of variation in responses, experiments should be run more than one time. This action is called replication. An experiment which is run two times has two replicates—replicate 1 is the first complete set of trials run and replicate 2 is the second group. Replication is not fulfilled by recording more than one response for an experimental trial. It requires distinct multiple runs of the experiment. Replication of designed experiments allows us to obtain an estimate of both central tendency and dispersion of the average level effects. It also allows us to estimate the amount of experiment error. The Western 2^2 array shown below is a full factorial with 2 replications.

tc	A	B	AB	Run 1	Run 2	Average Response
(1)	-1	-1	+1			R_1
a	+1	-1	-1			R_2
b	-1	+1	-1			R_3
ab	+1	+1	+1			R_4

The response for each treatment combination is filled in the "Run 1" column for the first replicate of the experiment, and in the "Run 2" column for the second replicate. These two results are used to calculate the average responses, R_1–R_4 for the experiment. The R_1–R_4 values are used to determine the magnitude of the average level effects in the same fashion that we studied earlier. For example, the partially completed array might appear as follows after the first run of an experiment with two replicates:

tc	A	B	AB	Run 1	Run 2	Average Response
(1)	-1	-1	+1	63		R_1
a	+1	-1	-1	56		R_2
b	-1	+1	-1	58		R_3
ab	+1	+1	+1	62		R_4

All four trials (or treatment combinations) would be repeated again and the results filled in the column labeled "Run 2."

tc	A	B	AB	Run 1	Run 2	Average Response
(1)	-1	-1	+1	63	65	R_1
a	+1	-1	-1	56	59	R_2
b	-1	+1	-1	58	54	R_3
ab	+1	+1	+1	62	60	R_4

The average responses are simply the arithmetic means of both the Run 1 and Run 2 results, as shown in the following set of equations:

$$R_1 = \frac{(63 + 65)}{2} = 64$$

$$R_2 = \frac{(56 + 59)}{2} = 57.5$$

$$R_3 = \frac{(58 + 54)}{2} = 56 \qquad (13\text{-}2)$$

$$R_4 = \frac{(62 + 60)}{2} = 61$$

Placing these values in our array, yields the following:

tc	A	B	AB	Run 1	Run 2	Average Response
(1)	-1	-1	+1	63	65	64
a	+1	-1	-1	56	59	57.5
b	-1	+1	-1	58	54	56
ab	+1	+1	+1	62	60	61

Response tables and graphs are produced from the "Average Response" column. We shall leave the Run 1 and Run 2 columns in our array for now. In later chapters we shall examine a technique which utilizes these results to quantify variation. Often, one or more quality characteristics we are studying are classified as nominal-the-best and reduction of variation is our goal. Obviously, as the number of replications increase, we can estimate the central tendencies and variation of our average effects with increased accuracy. In industrial settings, however, we rarely have the opportunity to run many replicates. Experimentation costs money. It also jeopardizes production schedules and shipments. In spite of this, we must always run a minimum of two replicates. One replicate could produce responses which resulted from assignable cause variation—repeating the experiment helps minimize the probability of this occurring.

Randomization

Thus far, all the arrays we have examined have had the trials listed in a specific order. Statisticians always recommend that experimentation not be performed in the trial order listed. Performing experiments in a set order increases the risk of erroneous results due to noise factors such as ambient temperature, process changes, machine warm-up, or a variety of other noise factors that are often unrecognized by the experimenter.

For example, consider the array which follows. It represents an experiment which involves two control factors A and B, and the AB interaction. Unknown to the experimenters, both factors are significantly more reactive with very slight increases in temperature. That is, both will have significantly larger effects on the quality characteristic response as temperature increases.

Suppose the experiment is initiated on a Monday, and the first two treatment combinations are completed on that day. The ambient temperature is 80° as illustrated in the table which follows. The third and fourth treatment combinations are run the following day. The ambient temperature is 70°. What effect would the temperature change have on our experimental responses?

		tc	A	B	AB
day 1	80°	(1)	-1	-1	+1
	80°	a	+1	-1	-1
day 2	70°	b	-1	+1	-1
	70°	ab	+1	+1	+1

Studying the array and the ambient temperatures for each day, we can calculate the average temperature at each factor level in the same manner we calculate the average level effects for control factors and interactions. The calculations are shown as Equations 13-3 and 13-4.

$$T_{A_{-1}} = \frac{(80° + 70°)}{2} = 75°$$

$$T_{A_{+1}} = \frac{(80° + 70°)}{2} = 75°$$

$$T_{B_{-1}} = \frac{(80° + 80°)}{2} = 80°$$

$$T_{B_{+1}} = \frac{(70° + 70°)}{2} = 70° \tag{13-3}$$

$$T_{AB_{-1}} = \frac{(80° + 70°)}{2} = 75°$$

$$T_{AB_{+1}} = \frac{(80° + 70°)}{2} = 75°$$

Using the results above, we can calculate the temperature differences for each control factor as follows:

$$T_{(A_{+1} - A_{-1})} = 75° - 75° = 0°$$

$$T_{(B_{+1} - B_{-1})} = 70° - 80° = -10° \tag{13-4}$$

$$T_{(AB_{+1} - AB_{-1})} = 75° - 75° = 0°$$

The average change in temperature for control factor B during this experiment has an absolute value of $10°$. Factor A and the AB interaction experienced no average change in temperature. Since we stated A and B were both more significant with temperature increases, we can see that B might appear significant in the analysis due to temperature alone. A variety of noise factors could change with time. This example illustrates why it is important to run experimental trials in random order.

Blocking

If we suspect unavoidable noise factors will influence our experimental results, we can choose to control which column (or columns) will be affected by a technique known as blocking. Consider the experimental array we have just examined in randomization once again.

		tc	A	B	AB
day 1	80°	(1)	-1	-1	+1
	80°	a	+1	-1	-1
day 2	70°	b	-1	+1	-1
	70°	ab	+1	+1	+1

If we knew, or even suspected, that noise factors were unavoidable and might influence the measured response in our experiment, we would have attempted to minimize their effect. We certainly do not want either control factor A or B to be affected—the experiment is designed to investigate their influence on the output characteristic response being studied. Since interactions are not always significant, we may be willing to sacrifice the AB interaction column.

Our goal is to arrange the order of the treatment combination runs in a fashion which maintains the integrity of A and B for noise factors which change levels from day 1 to day 2. Recall from our previous example that B was effected by temperature because both tc's with B at the -1 level were run on day 1, while both tc's with B at the +1 level were run on day 2. We must re-arrange the array to cause the AB interaction to be effected, while maintaining the independence of A and B. Consider the following array:

	tc	A	B	AB
day 1 — 80°	(1)	-1	-1	+1
day 1 — 80°	a	+1	-1	-1
day 2 — 70°	b	-1	+1	-1
day 2 — 70°	ab	+1	+1	+1

In the first treatment combination, (1), AB is at the +1 level and the temperature is 80°. If we rearrange the order of the array to include the other AB_{+1} run on day 1, it too will be run at 80°. This is illustrated below:

	tc	A	B	AB
day 1 — 80°	(1)	-1	-1	+1
day 1 — 80°	ab	+1	+1	+1
day 2 — 70°	b	-1	+1	-1
day 2 — 70°	a	+1	-1	-1

If we run treatment combination (1) first, and treatment combination *ab* second, they will both be subjected to the temperature and/or noise factors present during day 1. Running combinations *b* and *a* third and fourth will subject them to temperature and/or noise factors present during day 2, as illustrated in the following array.

		tc	A	B	AB
day 1	80°	(1)	-1	-1	+1
	80°	ab	+1	+1	+1
day 2	70°	b	-1	+1	-1
	70°	a	+1	-1	-1

The effect of rearranging the order of the experimental trials is illustrated by calculating the average temperatures for control factors A and B, and for the AB interaction. This is shown in Equations 13-5 and 13-6. Compare the results of these calculations with the earlier randomization example given in Equations 13-3 and 13-4. While we cannot control noise factors, we can arrange our experiment to control or at least direct their influence.

$$T_{A_{-1}} = \frac{(80° + 70°)}{2} = 75°$$

$$T_{A_{+1}} = \frac{(80° + 70°)}{2} = 75°$$

$$T_{B_{-1}} = \frac{(80° + 70°)}{2} = 75°$$

$$T_{B_{+1}} = \frac{(80° + 70°)}{2} = 75° \tag{13-5}$$

$$T_{AB_{-1}} = \frac{(70° + 70°)}{2} = 70°$$

$$T_{AB_{+1}} = \frac{(80° + 80°)}{2} = 80°$$

$$T_{(A_{+1} - A_{-1})} = 75° - 75° = 0°$$

$$T_{(B_{+1} - B_{-1})} = 75° - 75° = 0°$$

$$(13-6)$$

$$T_{(AB_{+1} - AB_{-1})} = 80° - 70° = 10°$$

Rearranging the order in which the experimental trials are run has shifted the noise factor effect to the AB interaction column. This does not mean the control factors were not subjected to noise - both A and B were run at both 80° and 70°. We merely caused the -1 and +1 levels of the control factors to be subjected to the same range of noise factor variation present from day 1 to day 2.

Any analysis of the AB interaction column must now be performed with caution. By design, we are analyzing the AB interaction *and* the noise factor effects concurrently. If the AB interaction appears significant, we must acknowledge the significance may be due to the noise factor variation. Likewise, if the AB interaction appears insignificant in the analysis of effects on the characteristic response, we must once again acknowledge the presence of noise factors and question if they tended to minimize the true effect of the AB interaction.

In the example presented, the AB interaction was used to rearrange the order of the experimental trials. The AB interaction was used as the *block generator* for determining the order of the trials. Blocking is sometimes represented as follows:

AB = Block Generator

(+) Block

A	B
-1	-1
+1	+1

(-) Block

A	B
-1	+1
+1	-1

The (+) Block and (-) Block headings refer to the level of the AB interaction. This notation gives the same information as the experimental array with the trials rearranged earlier. Compare the notation above with the array we constructed:

tc	A	B	AB
(1)	-1	-1	+1
ab	+1	+1	+1
b	-1	+1	-1
a	+1	-1	-1

In larger arrays, it is possible to block the experiment by using one or more interaction columns. The column(s) chosen for blocking should always be the one(s) you feel are least significant. Additional information can be found in some of the references at the end of this chapter.

Example 13.1: Bearing Diameter

The manufacture of a split bearing from steel coil may be accomplished with a variety of manufacturing methods. In general terms, the process consists of cutting a length of coil to obtain the volume of material required, pre-forming to obtain a semi-cylindrical form, and final forming to establish the dimensions required by the engineering specification. The end result is a product as illustrated in Figure 13-1.

Figure 13-1: Bearing Configuration

A manufacturer investigating the final forming process selected three control factors for evaluation in a full factorial experimental array. These factors were the length of the blank cut from the coil, the forming tool for the inside diameter, and the forming tool for the outside diameter. While multiple quality characteristic responses were selected for evaluation, we shall limit our discussion in this example to the outside diameter of the bearing.

To understand the recorded response values, a brief description of the outside diameter (OD) measurement method is needed. In actual service, this type of bearing is pressed into a housing and requires a friction fit. If the bearing is too large, it will either not install in the housing at all, or it will become distorted during the installation process. If the bearing is too small, it will install in the housing loosely, and will not remain in place under load. Bearing OD is quantified by placing a bearing between calibrated gage blocks, applying a specified force, and measuring the deflection of the upper block. An illustration of the checking device is shown below. The OD responses in this example are all coded values with $1 = 0.0001$ inches.

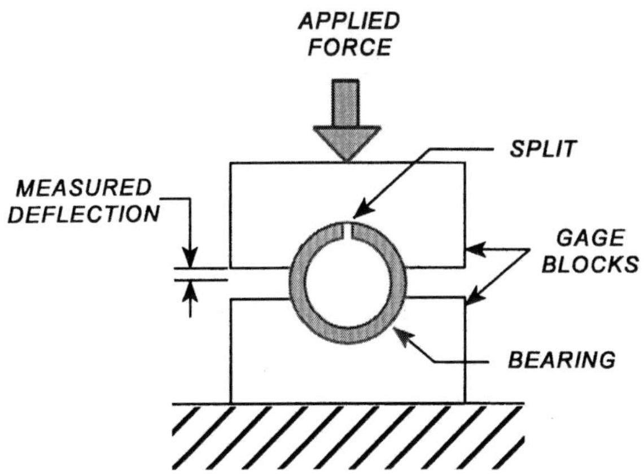

Figure 13-2: Bearing OD Measurement

Within the range of normal operating parameters, high and low levels were established for the inside and outside diameter forming tools, and for the material blank length. The factors were placed in a Taguchi L_8 experimental array, with four columns left empty to evaluate interaction effects between the control factors. The L_8, like many Taguchi experimental arrays, does contain columns which may be used for interactions. The format for a few Taguchi experimental arrays is shown in the Appendix, "Experimental Arrays and Response Tables." (Columns are labeled at the bottom with letter codes denoting either the control factor, or the interaction applicable to the column.) The L_8 array and the average measured responses for each trial are shown below in Figure 13-3. The average response was obtained from 40 measurements per trial.

T R I A L	Control Factors							Results
	A Blank Length	B ID Tool	AB	C OD Tool	AC	BC	ABC	Average Response
1	short	small	1	small	1	1	1	-10.2
2	short	small	1	large	2	2	2	-8.5
3	short	large	2	small	1	2	2	-11.5
4	short	large	2	large	2	1	1	-9.4
5	long	small	2	small	2	1	2	-9.6
6	long	small	2	large	1	2	1	-7.9
7	long	large	1	small	2	2	1	-10.9
8	long	large	1	large	1	1	2	-8.9

Figure 13-3: L_8 Experimental Array

The average response for the experiment is calculated as shown in Equation 13-7.

$$\overline{T} = \frac{\sum\limits_{i=1}^{n} R_i}{n} = \frac{-10.2 - 8.5 - 11.5 - 9.4 - 9.6 - 7.9 - 10.9 - 8.9}{8} \tag{13-7}$$

$$\overline{T} = -9.6125 \approx 9.6$$

The equations for calculating the average level effect for each control factor and interaction are shown below:

$$A_1 = (\,-10.2 - 8.5 - 11.5 - 9.4\,) \,/\, 4 = -9.9$$

$$A_2 = (\,-9.6 - 7.9 - 10.9 - 8.9\,) \,/\, 4 = -9.325$$

$$B_1 = (\,-10.2 - 8.5 - 9.6 - 7.9\,) \,/\, 4 = -9.05$$

$$B_2 = (\,-11.5 - 9.4 - 10.9 - 8.9\,) \,/\, 4 = -10.175$$

$$(AB)_1 = (\,-10.2 - 8.5 - 10.9 - 8.9\,) \,/\, 4 = -9.625$$

$$(AB)_2 = (\,-11.5 - 9.4 - 9.6 - 7.9\,) \,/\, 4 = -9.6$$

$$C_1 = (\,-10.2 - 11.5 - 9.6 - 10.9\,) \,/\, 4 = -10.55$$

$$C_2 = (\,-8.5 - 9.4 - 7.9 - 8.9\,) \,/\, 4 = -8.675 \tag{13-8}$$

$$(AC)_1 = (\,-10.2 - 11.5 - 7.9 - 8.9\,) \,/\, 4 = -9.625$$

$$(AC)_2 = (\,-8.5 - 9.4 - 9.6 - 10.9\,) \,/\, 4 = -9.6$$

$$(BC)_1 = (\,-10.2 - 9.4 - 9.6 - 8.9\,) \,/\, 4 = -9.525$$

$$(BC)_2 = (\,-8.5 - 11.5 - 7.9 - 10.9\,) \,/\, 4 = -9.7$$

$$(ABC)_1 = (\,-10.2 - 9.4 - 7.9 - 10.9\,) \,/\, 4 = -9.6$$

$$(ABC)_2 = (\,-8.5 - 11.5 - 9.6 - 8.9\,) \,/\, 4 = -9.625$$

These results are used to produce the response table in Figure 13-4 and the response graph shown in Figure 13-5. Differences are calculated with full precision. Since 1=0.0001 inches, however, results are rounded to one decimal place on the response graph axis.

	A Blank Length	B ID Tool	AB	C OD Tool	AC	BC	ABC
Level 1 Average	-9.9	-9.05	-9.625	-10.55	-9.625	-9.525	-9.6
Level 2 Average	-9.325	-10.175	-9.6	-8.675	-9.6	-9.7	-9.625
Difference	-.575	1.125	-.025	-1.875	-.025	.175	.025

Figure 13-4: Response Table for Bearing Diameter

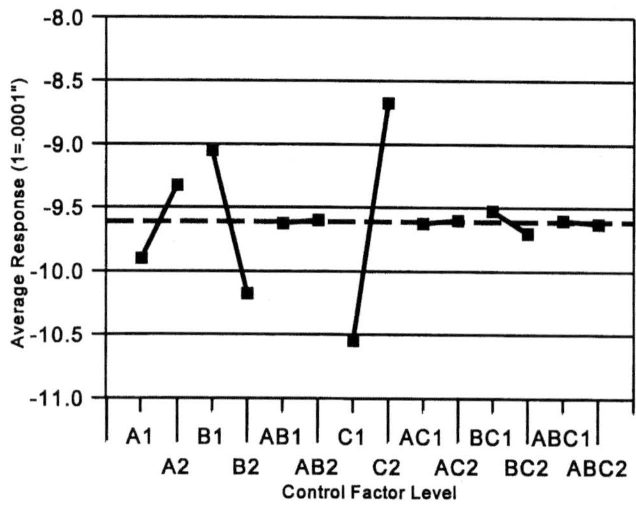

Figure 13-5: Response Graph for Bearing Diameter

The response graph for bearing diameter illustrates the effects of the control factors and their interactions on the quality characteristic measured. While there are both graphical and computational statistical methods available to analyze response graphs, they are beyond the scope of this text. We shall rely on visual interpretation. This method, coupled with process knowledge, is normally adequate for interpretation of results and process optimization within the bounds of the experiment.

For this experiment, the graph indicates that all of the interaction effects, AB, AC, BC, and ABC have no significant effect on the bearing outside diameter. In fact, these factors represent the magnitude of experimental error.

It is apparent that the most significant control factor for altering outside diameter is factor C, the OD tool. Using the large OD tool, denoted by C2 in the response graph, will result in a larger outside diameter. And, of course, using the small one will result in a smaller outside diameter.

In terms of magnitude of effect, the OD tool is followed by factor B, the ID tool. This control factor's effect has the opposite effect on the bearing's outside diameter. Changing to a small OD tool appears to cause the measured OD to become larger. While this does not "make sense" at first glance, recall the measurement method illustrated in Figure 13-3. The outside diameter is quantified by deflection of the gage block when the bearing is placed under load. Since a smaller ID tool could reasonably be expected to result in a smaller ID, this will increase the wall thickness of the bearing. The increased wall thickness, in turn, will make the bearing more resistant to deformation under load and the measured gage block deflection will be smaller. This will increase the force required to install the bearing in a housing and cause higher retention forces.

The final factor, A, is the cut blank length of material. This control factor response illustrates the value of process knowledge and experience. Even though it appears to be significant, it may or may not have practical value in targeting the OD of the bearing studied. The reason for this lies with the international standard tolerance for the part. The width of the specification is 24 (once again, 1=0.0001"). Since changing the level of factor A will cause a change of approximately 0.6 in the diameter of the bearing, this amounts to only 2.4% of the tolerance.

        This information is sufficient for now, however, we shall examine this example in more detail in later chapters. In Chapter 15, we will examine how to use information presented in a response graph to obtain a desired average output. Near the end of this text, we will examine how experimental results can be used to create a robust process.

References

1. American Supplier Institute *Taguchi Methods, Introduction to Quality Engineering Course Manual,* Version 2.1, American Supplier Institute, Dearborn, MI, 1991.

2. Barker, T. B. *Engineering Quality by Design—Interpreting the Taguchi Approach*, Marcel Dekker, New York, 1990.

3. Barker, T. B. *Quality by Experimental Design,* 2nd edition, Marcel Dekker, New York, 1994.

4. Berger, R.W (ed.) and Pyzdek, T. (ed.) *Quality Engineering Handbook*, ASQC Quality Press, Milwaukee, WI, and Marcel Dekker, New York, 1992.

5. Lochner, R.H. and Matar, J.E. *Designing for Quality—An Introduction to the Best of Taguchi and Western Methods of Statistical Experimental Design*, Quality Resources, New York and ASQC Quality Press, 1990.

6. Montgomery, D.C. and Myers, R.H. *Response Surface Methodology —Process and Product Optimization Using Designed Experiments,* John Wiley & Sons, New York, 1995.

7. Taguchi, G. *System of Experimental Design,* UNIPUB/Kraus International Publications, White Plains, NY, and American Supplier Institute, Dearborn, MI, 1987.

8. Taguchi, G. *Introduction to Quality Engineering*, Asian Productivity Organization, Tokyo, 1986.

14

Fractional Factorial Arrays

In this chapter we shall explore an approach to experimentation which does not always utilize the interaction columns of our arrays to evaluate control factor interactions. As we shall see, these columns are used instead to evaluate additional control factors. This greatly reduces the number or experimental trials needed in an investigation. This reduction may at times determine whether or not an experiment is even performed in a manufacturing environment, where time and productivity are crucial. The advantage does not come without cost for control factors which are placed in interaction columns become confounded with interactions, should they exist. This approach to experimental design involves the use of fractional factorial arrays.

It must be noted that while fractional factorial experimentation has increased in popularity in the past 10 or 15 years, it is often a source of heated debate. Much of the increase is due to the work of Dr. Genichi Taguchi, who has published a comprehensive series of arrays and an approach to experimentation that has been warmly received by some in industry, particularly it seems, by engineers. Statisticians, however, generally favor the more "scientific" or determinative approach resulting from the use of full factorial experimentation. The truth is, like most situations in a manufacturing environment, there are advantages and disadvantages to both methods. We shall explore the utility of both approaches, and discuss the limitations and applicability of each.

Fractional Factorial Arrays

For our purposes, a suitable definition is as follows:

> *A fractional factorial array is one which attempts to analyze one or more control factors in interaction columns.*

In fractional factorials, one or all of the interaction columns may be used to examine the effects of additional control factors. The number of trials needed for a fractional factorial experiment is:

$$treatment\ combinations\ =\ tc\ =\ 2^{k-p}$$

$$where:\ k\ =\ number\ of\ factors \tag{14-1}$$
$$p\ =\ fractionalizing\ element$$

The variable p is called the fractionalizing element. By subtracting this value from the number of factors in our experiment, as shown in Equation 14-1, it determines the fraction of full factorial treatment combinations we desire to run. It determines the fraction of full factorial trials according to the following relationship:

$$Fraction\ desired\ =\ (\tfrac{1}{2})^{p} \tag{14-2}$$

For example, suppose we design an experiment which will evaluate seven main effects, A B C D E F and G. The value of k is seven and a full factorial design would require 2^{7} or 128 trials. If we have time and cost limitations which preclude more than 8 trials, we are able to run only 1/16 of the trials required for a full factorial design (8/128=1/16). The value of the fractionalizing element, p, is therefore equal to four.

This relationship may be calculated from Equation 14-2, with $p=4$, as shown below:

$$Fraction\ desired\ = \left(\tfrac{1}{2}\right)^{p}$$

$$Fraction\ desired\ = \left(\tfrac{1}{2}\right)^{4} \tag{14-3}$$

$$Fraction\ desired\ = \left(\tfrac{1}{16}\right)$$

The table shown below, illustrates the number of treatment combinations required by a few simple two-level experimental arrays, for various values of the fractionalizing element, p. (The first row, with $p=0$, corresponds to the full factorial design.)

Table 14-1: Number of Trials for Various Fractionalizing Elements

		2^k Experiment Design						
p	fraction	2^2	2^3	2^4	2^5	2^6	2^7	2^8
0	1	4	8	16	32	64	128	256
1	1/2		4	8	16	32	64	128
2	1/4				8	16	32	64
3	1/8					8	16	32
4	1/16						8	16

The shaded region of the table above indicates the limit of fractionalization. Even if all interaction columns are used to evaluate control factors our designs will require one trial more than the number of control factors being evaluated. For example, we could evaluate no more than 7 control factors in 8 trials. The fractionalizing element, p, has other significance which will be explored in the next two sections.

Confounding

Let's return to a full factorial design to evaluate two control factors, A and B. The array illustrating one of the experimental designs we have been working with thus far is shown below. Examine the interaction column for a moment—what if we felt very confident the interaction does not exist? Could we do anything else with this column? The answer is yes; we can insert an additional control factor in the experimental array, provided we establish the levels of the control factor exactly as shown in the AB interaction column.

tc	A	B	AB
(1)	-1	-1	+1
a	+1	-1	-1
b	-1	+1	-1
ab	+1	+1	+1

The new array is a fractional factorial design. It specifically excludes the analysis of an interaction column from the full factorial array. This fractional factorial design is shown below.

tc	A	B	C
(1)	-1	-1	+1
a	+1	-1	-1
b	-1	+1	-1
ab	+1	+1	+1

This array has the ability to analyze three main effects in four treatment combinations or trials. It was selected with the intent to establish the optimum or preferred control factor levels out of the 2^3 possible control factor combinations. There is, however, a potential problem with this array. If the AB interaction we predetermined to be insignificant is important, main effect C will also appear important in the response table and response graph, even though C itself may be insignificant. Since the AB column and the C column are exactly the same, we cannot determine in the analysis of output whether the AB interaction or control factor C is responsible for any observed effect on the output characteristics being studied. We characterize this confusion by saying that the *AB interaction and C main effect are confounded.*

The fractional factorial version of the array is a 2^{3-1} design. The value of p is 1, and the number of runs is $(1/2)^1$ times the number required for a full factorial with 3 control factors. In the response analysis of this array, the AB interaction column is equivalent to the C main effect column, as shown in Equation 14-4.

$$C \equiv AB$$

$$(C \text{ confounded with } AB)$$

(14-4)

This notation does not mean that the interaction effect is equivalent to the C main effect. It merely indicates the factor levels in their respective columns are the same and the two are confounded.

Since $C \equiv AB$ makes this array design possible, $C \equiv AB$ is called the *design generator* for this array. When determining the amount of confounding in a fractional factorial array, there are p generators.

We know the C main effect is confounded with the AB interaction. They both occupy the same column of our array, and therefore their levels are exactly the same. Since we now have an experimental array with three main effects, we should at least examine any potential interactions between control factors A and C, and any between B and C.

To examine what an AC interaction would look like in terms of levels, consider the array which follows. For illustrative purposes, we have multiplied the A and C columns together and listed the resulting interaction levels in the extra column with the heading AC.

tc	A	B	C	AC
(1)	-1	-1	+1	-1
a	+1	-1	-1	-1
b	-1	+1	-1	+1
ab	+1	+1	+1	+1

If we examine the AC column, we can see it is exactly the same as the column for the B main effect. We cannot add an additional column to the array for the interaction since the column already exists for factor B. Therefore, in any analysis of the response for this fractional factorial array, the main effect of control factor B will be confounded with the AC interaction. This is illustrated by the two unshaded columns in the array, as shown below:

tc	A	B	C	AC
(1)	-1	-1	+1	-1
a	+1	-1	-1	-1
b	-1	+1	-1	+1
ab	+1	+1	+1	+1

This represents a possible problem for us. We knew that we were confounding factor C's effect with the AB interaction when we designed the array. Now, we see that the effect of control factor B is confounded with the effect of the AC interaction. A response table or graph which indicates B is important in controlling the output response must be viewed with caution. Any effect on the response that appears to originate with the levels of B might be due to an AC interaction.

What would a BC interaction look like? To examine this, we have removed the AC column and added a BC column to the right of our array in the illustration below. Once again, we find the BC interaction column by multiplying the levels from columns B and C. This time, we find the BC interaction has exactly the same levels as the column for control factor A, as shown below. This experimental design causes the main effect A to become confounded with the BC interaction effect.

tc	A	B	C	BC
(1)	-1	-1	+1	-1
a	+1	-1	-1	+1
b	-1	+1	-1	-1
ab	+1	+1	+1	+1

We initially felt the AB interaction was not significant and placed control factor C in the third column. While that seemed simple enough, we now have the experimental array which appears below:

tc	A ≡ BC	B ≡ AC	C ≡ AB
(1)	-1	-1	+1
a	+1	-1	-1
b	-1	+1	-1
ab	+1	+1	+1

In order to get valid response analysis for any of the control factors A, B, or C, all the interactions AB, AC, and BC must be insignificant. Because we placed factor C in the AB interaction column, this array now has all main effects confounded with first order (two-factor) interactions.

Design Resolution

Multiplying individual columns to determine confounding is accurate, but time consuming (especially with larger arrays). There is an easier method, as illustrated below:

1. Specify the design generator(s).

    $C \equiv AB$

2. Multiply both sides of each design generator by the control factor(s) being substituted for the interaction.

    $CC \equiv CAB$

3. Replace any $(factors)^2$ with the identity symbol, I. List all combinations of factors after the identity symbol.

    $I = ABC$

4. Count the number of algebraic terms (I is a term). There will be 2^P terms in the defining contrast. If you do not obtain 2^P terms, multiply the terms by one another to obtain the missing terms.

    $I = ABC$

    The resulting statement with 2^P terms is known as the *defining contrast* of the experiment.

5. Count the number of factors in the shortest term (not I). This is the *design resolution* of the experimental array.

    $I = ABC \Rightarrow 3$ factors in shortest term $\Rightarrow$ *resolution III*

6. Multiply the defining contrast by each control factor
   successively to determine confounding.

$$AI = AABC \Rightarrow A = A^2BC \Rightarrow A = IBC \Rightarrow A \equiv BC$$

$$BI = BABC \Rightarrow B = AB^2C \Rightarrow B = AIC \Rightarrow B \equiv AC$$

$$CI = CABC \Rightarrow C = ABC^2 \Rightarrow C = ABI \Rightarrow C \equiv AB$$

The design resolution of arrays indicates the level of confounding as
follows:

Resolution III - Main effects of the array are not
confounded with one another. They are confounded with
two- factor interactions.

Resolution IV - Main effects are not confounded with one
another. Main effects are not confounded with two-factor
interactions. Two-factor interactions are confounded with
one another.

Resolution V - Neither main effects nor two-factor
interactions are confounded with one another. Two-factor
interactions are confounded with three-factor interactions.

Example 14.1: Confounding

Consider the following 2^3 full factorial array which requires eight
treatment combinations. The array evaluates three main effects, three
two-factor interactions, and one three-factor interaction. If we felt
confident some of the interactions were insignificant, and we had a desire
to evaluate additional control factors, we could substitute the additional
control factors in interaction columns we felt were least significant.

tc	A	B	C	AB	AC	BC	ABC
(1)	-1	-1	-1	+1	+1	+1	-1
a	+1	-1	-1	-1	-1	+1	+1
b	-1	+1	-1	-1	+1	-1	+1
ab	+1	+1	-1	+1	-1	-1	-1
c	-1	-1	+1	+1	-1	-1	+1
ac	+1	-1	+1	-1	+1	-1	-1
bc	-1	+1	+1	-1	-1	+1	-1
abc	+1	+1	+1	+1	+1	+1	+1

For example, assume we want to evaluate some quality characteristic response by varying five control factors, A B C D and E. Normally, an experiment with five control factors would require 2^5 or 32 treatment combinations for a full factorial array. If we desire to limit our array to eight tc's to minimize cost and downtime, we must run a fractional factorial design. Since the ABC interaction is most likely to be insignificant, we would substitute one of the additional control factors in the ABC interaction column. The remaining control factor would replace one of the three two-factor interactions. We choose the two-factor interaction we believe least significant and place the control factor in that column.

Suppose that we have enough experience with the process to identify which interaction columns are most likely to be insignificant to the output characteristic we are studying. We decide to place control factor E in the ABC interaction column, and control factor D in the AC interaction column. The fractional factorial array with these substitutions is shown on the following page.

tc	A	B	C	AB	D	BC	E
(1)	-1	-1	-1	+1	+1	+1	-1
a	+1	-1	-1	-1	-1	+1	+1
b	-1	+1	-1	-1	+1	-1	+1
ab	+1	+1	-1	+1	-1	-1	-1
c	-1	-1	+1	+1	-1	-1	+1
ac	+1	-1	+1	-1	+1	-1	-1
bc	-1	+1	+1	-1	-1	+1	-1
abc	+1	+1	+1	+1	+1	+1	+1

Since we have taken a 2^5 full factorial design and reduced it to 2^3 treatment combinations, we know the value of p is 2. Our fractional factorial is a 2^{5-2} design. We also know there are 2 design generators. They are:

$$D \equiv AC \qquad E \equiv ABC$$

Multiplying both design generators by the control factors on the left sides of their respective equations we obtain:

$$I = ACD \qquad\qquad I = ABCE$$

This yields:

$$I = ACD, ABCE$$

There are only 3 terms above—I, ACD, and ABCE. We know the defining contrast must have 2^p or 2^2 terms for this experimental array. The missing term is found by multiplying the terms with factors together:

$$ACD \times ABCE = BDE$$

The defining contrast for our fractional factorial array is:

$$I = ACD, BDE, ABCE$$

Since the shortest term has three factors, the design resolution of this experiment is III. As we already know from the placement of factors D and E, this resolution indicates main effects will not be confounded with one another, but they will be confounded with two-factor and higher interactions. The complete confounding for this array is determined as follows:

__I = ACD__	__I = BDE__	__I = ABCE__
A ≡ CD	B ≡ DE	A ≡ BCE ⇒ AB ≡ CE
C ≡ AD	D ≡ BE	⇒ AC ≡ BE
D ≡ AC	E ≡ BD	⇒ AE ≡ BC
		B ≡ ACE
		C ≡ ABE
		E ≡ ABC

Organizing this information in terms of the control factors and interactions used as column headings in our array, we obtain:

$$A \equiv CD \equiv BCE$$
$$B \equiv DE \equiv ACE$$
$$C \equiv AD \equiv ABE$$
$$D \equiv AC \equiv BE$$
$$E \equiv BD \equiv ABC$$
$$AB \equiv CE$$
$$BC \equiv AE$$

This details the confounding present in the 2^{5-2} experimental array, when control factor D is placed in the AC interaction column and control factor E is placed in the ABC interaction column. The confounding can be verified by multiplying the appropriate columns together as illustrated with the ABC array earlier in this unit.

As we can see, even though we identified the interaction columns for AC and ABC as insignificant, we created a much greater amount of confounding in the experimental array when we substituted control factors D and E for these interaction columns. In the analysis of a response for a fractional factorial array we must be aware, that confounding is present to a much greater degree than simple substitution might appear to indicate.

Screening Experiments

With the amount of confounding between main effects and first and second order interactions, many of us might decide to use only full factorial arrays. This may be a valid choice, depending upon the nature of the process under investigation and the amount of time and money available for experimentation. Full factorial arrays are especially useful if the goal of the experiment is research or "complete" process knowledge.

Fractional factorials, however, have distinct advantages with respect to the amount of expenditure necessary to investigate many factors. For example, a 2^3 full factorial array will investigate three control factors in eight treatment combinations. If all interaction columns are used for control factors, the same array can be used for seven control factors in a 2^{7-4} fractional factorial experiment. The seven control factor full factorial array would have required 2^7 or 128 treatment combinations. If the interactions are insignificant, or if total process knowledge is not the desired goal of the experimental effort, the amount of downtime and experimental cost for the seven factor array is reduced 93.75% by utilizing the 2^{7-4} array. This is often the type of choice we are faced with in manufacturing environments. Downtime and cost usually determine if we perform experiments in a production environment.

As we stated in an earlier chapter, second order (three-factor) interactions are seldom significant in a manufacturing process. First order interactions (two-factor), while they may be significant at times, are usually not unless at least one of the control factors in the interaction is significant.

While a fractional factorial experiment will not provide the same level of clarity possible from a full factorial, it will often provide insight into which control factors might be important with respect to one or more quality characteristic responses. More often the factors brainstormed in a manufacturing environment as control factors are not all significant in determining process outputs. A fractional factorial allows us to investigate many factors with minimal expense and identify which ones might be important to process control. One argument used in favor of fractional factorial arrays is represented simplistically in Figure 14-1.

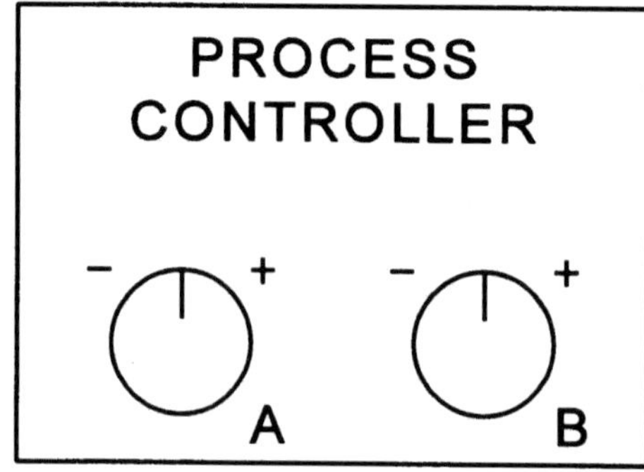

Figure 14-1: Process Control Panel

As illustrated, we have the ability to adjust control factor A. We also have the ability to adjust control factor B. There is no knob, button, or dial for the operator to control AxB. It may be better to investigate the factors the operator can directly control.

The type of experimental array chosen will be largely dependent upon the specific application. Often in industrial situations, DOE techniques are not applied until one or more factors effect a process in an adverse manner. Scrap increases, downtime increases, and production schedules are altered. Shipments and customer satisfaction are in jeopardy. In the midst of all this confusion, and with the possible threat of shutting down a customer's operation, there is seldom time to perform all of the treatment combinations necessary for full factorial arrays. A fractional factorial is often the best choice to evaluate many control factors, it requires minimal effort and returns the process to a viable production mode. Fractional factorials are often used as screening experiments to identify the critical control factors. After the critical factors have been identified, follow up experimentation can be accomplished with full factorials if the expenditure can be justified. The effectiveness of the fractional factorial array is such that process improvements are almost always realized upon response analysis. While they do not necessarily provide a complete knowledge of control factor ⇒ quality characteristic relationships, the improvements obtained are often great enough to make additional experimentation unnecessary.

Example 14.2: Coil Profile

In this example, a manufacturing facility was sintering bronze particles onto the surface of copper-coated steel coils. One of the manufacturer's goals was to minimize the variation in thickness on finished product. Through brainstorming sessions with line operators, the process engineer and department manager both identified coil profile as a key input to the sintering operation. Coil profile is the term used to describe the flatness of the steel across its width. The magnitude of deviation from a flat state is called crossbow, as illustrated in Figure 14-2.

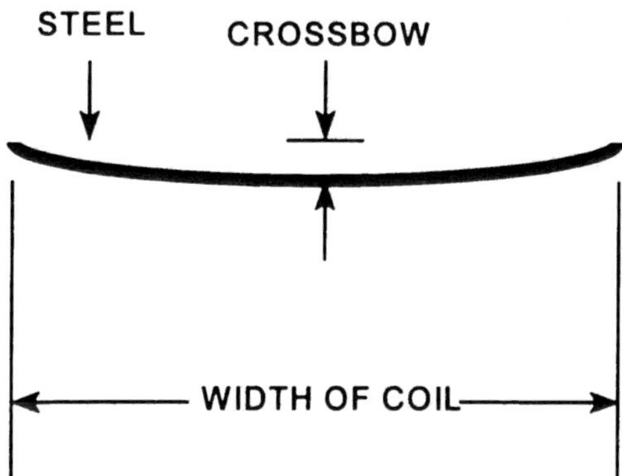

Figure 14-2: Cross-Section of Steel Coil

Crossbow has an adverse affect on thickness control of sintered product because of the increased volume of bronze powder deposited and leveled on the middle of the coil. Once sintered to the steel, this area of the coil becomes thicker than the two ends. The process consists of unwinding a steel coil, passing it through a coil flattener, depositing the bronze powder on the top surface of the coil, and sintering the bronze in a continuous oven. The process flow is shown in Figure 14-3.

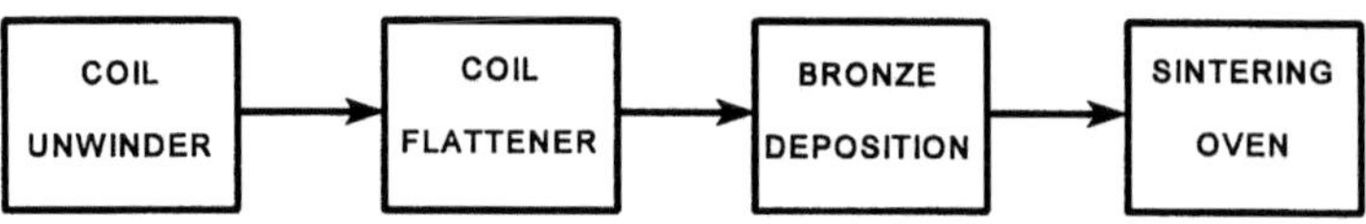

Figure 14-3: Process Flow Diagram

From the list of factors theorized to be important with respect to crossbow, three were selected for experimentation. These factors, and their respective levels, are shown in the following table.

Factor	Level 1	Level 2
Flattener Gap	low	high
Line Speed	slow	fast
Steel Supplier	1	2

The factors levels are considered proprietary by the manufacturer. The factors themselves are self-explanatory, except perhaps for the first one labeled flattener gap. This factor is a measure of the gap between the coil flattener rolls when the steel coil is processed through the flattener. Because of production requirements, the experiment was designed to run on production equipment in the shortest possible time. This necessitated placing the control factors in a 2^{3-1} fractional factorial array. The Taguchi L_4 design was selected, as shown in the following illustration.

Trial	Flattener Gap	Line Speed	Steel Supplier	Average Response
1	low	slow	1	.00035
2	low	fast	2	.00051
3	high	slow	2	.00037
4	high	fast	1	.00025

Experiments were run in random order, with two replications. Five samples were evaluated for crossbow magnitude from each trial of both replications. The average responses for each trial are shown in the array above. (Normal production historically averaged $0.0004 \pm 0.0001"$.) The average responses were then placed in a response table and graphed, as shown in Figures 14-4 and 14-5.

	Flattener Gap	Line Speed	Steel Supplier
Level 1	.00043	.00036	.00030
Level 2	.00031	.00038	.00044
Difference	-.00012	.00002	.00014

Figure 14-4: Response Table for Coil Profile

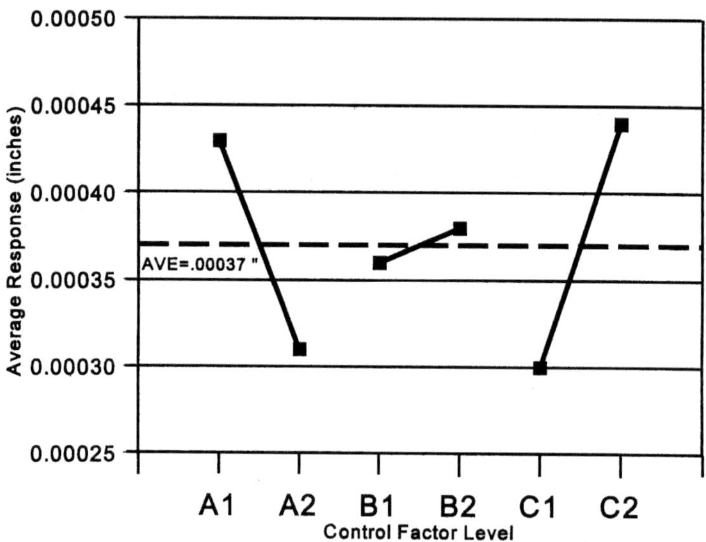

Figure 14-5: Response Graph for Coil Profile

The response graph seems to indicate control factors A and C have a larger effect on the measured output than factor B. We can not be certain, however, since all main effects are confounded with two-factor interactions for this fractional factorial array. It is possible, for instance, that the AC interaction (in the B column of this array) is equal in magnitude and opposite in sign to the B main effect. This condition would make control factor B appear unimportant, when in reality it might have a real effect on the measured output.

The department manager and the process engineer decided to run factor A, flattener gap, at level 2 to minimize the crossbow condition. They decided to source steel only from supplier 1, also to minimize the crossbow condition. Since line speed did not seem to have a significant effect on the measured quality characteristic output within the range of this experiment, they decided to run the process at the higher line speed to maximize productivity.

In the following chapter we shall examine how we can analyze the response graph and actually predict the output characteristic measurement when one or more control factor levels are selected. For this experiment, the manufacturer predicted an average crossbow response of 0.00024 inches. The factor levels were set and the process was run to verify the predicted response. After verification, these levels became standard operating procedures for the process. For further verification, a total of 57 steel coils were examined nine months after completion of this experiment. The average crossbow on these coils was 0.00025 inches. This result differed from the experiment's predicted average by only 10 millionth of an inch .

Although it certainly would have been possible, no additional experiments were performed on this process to further minimize crossbow. The results obtained were sufficient to insure the successful manufacture of product in all subsequent operations, making further expenditure for crossbow reduction undesirable.

Example 14.3: Liner Bonding

A manufacturer of pharmaceutical caps was experiencing a retention problem with tamper-evident liners designed to seal the top of a drug container. This is the foil or paper liner typically found on the top of vitamin or drug containers to verify the integrity of the product. The liner, or safety seal, is placed inside the cap during the manufacturing process and bonded to the cap liner with food-grade wax. During the filling process at the customer's facility, the top of the container is coated with an adhesive just prior to application of the cap. The wax is formulated to shear when the consumer removes the cap, while the adhesive is formulated to retain the safety seal on the container. The cross section of the cap assembly is illustrated in Figure 14-6.

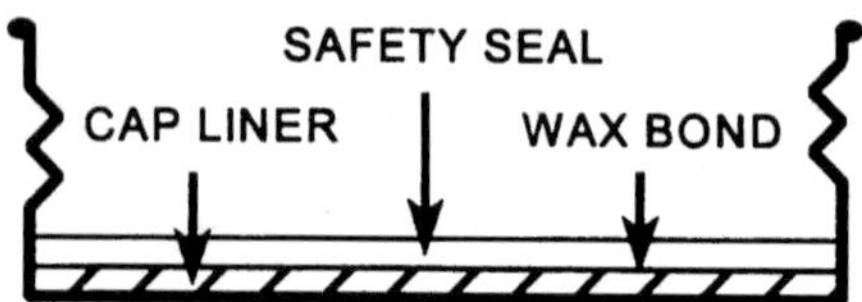

Figure 14-6: Cross Section of Cap with Safety Seal

The manufacture of the cap assembly required bonding the safety seal to the cap liner, cutting the bonded liner to the proper diameter, and inserting it into the cap shell. If the bond was made too strong, the safety seal would not shear from the cap liner when the consumer opened the container. Obviously, this condition had the potential to raise questions of product tampering in the mind of the consumer. On the other hand, if the bond between the safety seal and the cap liner was too weak, the safety seal would not be retained in the cap through cutting, installation, and transit to the pharmaceutical manufacturer. This also produced a missing seal condition and raised the question of product tampering in the mind of the consumer. The goal of the cap manufacturer was adequate retention to survive through the cap application process, without exceeding the strength of the adhesive designed to apply the safety seal to the top of the container. The caps were manufactured on multiple manufacturing lines, some with slightly different configurations. The manufacturer set out to examine the processes for common factors and determine their significance in the determination of liner bond strength.

The wax used to bond the two liners together was pre-applied to the cap liner. The liner bonding process consisted of two uncoilers to unwind rolls of both types of liner, a hot plate to apply heat and pressure to melt the wax and pre-bond the liners, and a final application roller to apply pressure and insure intimate contact between the two liners as the wax solidified. The bonded liners then progressed to the next station in a continuous flow where they were cut and inserted into the cap shell. A partial representation of the process is illustrated in Figure 14-7.

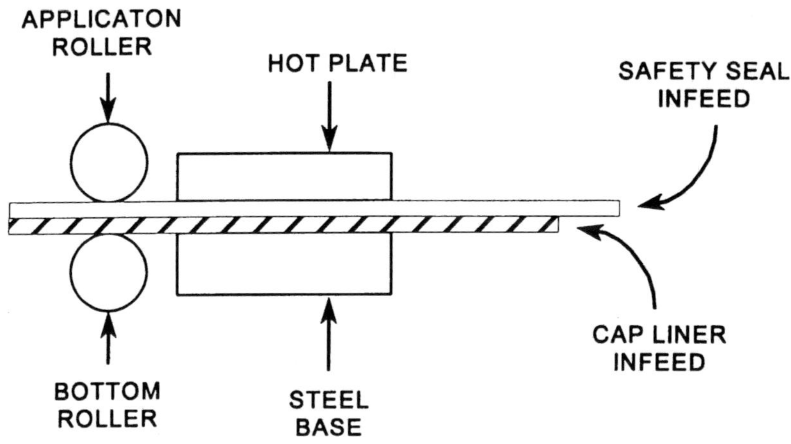

Figure 14-7: Liner Bonding Process

Manufacturing personnel brainstormed the poor adhesion problem and identified a number of potential control factors on a Pareto diagram. From the diagram they constructed, the following factors were selected for experimentation at the levels shown:

A. Roller surface finish: smooth (type 1), coarse (type 2)
B. Roller pressure (applied load): low, high
C. Wax type: type 1, type 2
D. Hot plate temperature (°F): 160°, 190°
E. Hot plate pressure (applied load, lb.): 0, 20
F. Safety seal type: old, new
G. Hot plate-to-roller distance: short, long

A Taguchi L_8 experimental array was selected for this experiment. All interaction columns were intended to be used for the evaluation of control factors. With the setup of the second experimental trial, personnel deemed factor G, the distance between the hot plate and the roller, to be too difficult to adjust. Because of severe time constraints while running this experiment in a manufacturing environment, the personnel involved decided not to attempt this adjustment and did not redesign the array to use the last column for any other factor. (Since this is the ABD interaction column in the notation shown below, they might have re-arranged the other factors or placed a different control factor here.)

Two complete replications were run in random order. Liner bond was evaluated by placing 200 completed cap assemblies in a rotating barrel tester to simulate severe transit vibration and purposely induce bond failure (200 was the capacity of the barrel). The number of caps completing the abrasion test intact was recorded in the results column of the array. The completed array and results are illustrated below.

TRIAL	Control Factors							Results		
	A Roller type	B Roller Load	C Wax Type	D Plate Temp	E Plate Load	F Liner Type	G	Run 1	Run 2	Ave.
1	1	low	1	160	0	old	-	7	93	50
2	1	low	1	190	20	new	-	109	192	150.5
3	1	high	2	160	0	new	-	85	128	106.5
4	1	high	2	190	20	old	-	154	180	167
5	2	low	2	160	20	old	-	43	70	56.5
6	2	low	2	190	0	new	-	142	127	134.5
7	2	high	1	160	20	new	-	94	52	73
8	2	high	1	190	0	old	-	140	160	150

Figure 14-8: L_8 Experimental Array

The response table for this experimental array is shown below. The ABD interaction effect has been placed in the table since factor G was not evaluated. The levels are from the Taguchi L_8 design.

	A Roller Type	B Roller Load	C Wax Type	D Plate Temp	E Plate Load	F Liner Type	ABD
Level 1	118.5	97.88	105.88	71.5	110.25	105.88	106.13
Level 2	103.5	124.13	116.13	150.5	111.75	116.13	115.88
Average Response Difference	15.00	26.25	10.25	79.00	1.50	10.25	9.75

The overall average response for the experiment is 111. The response graph for this experiment is illustrated in Figure 14-9.

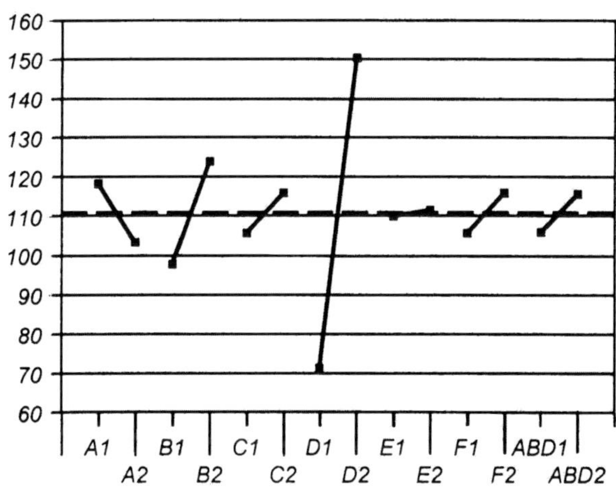

Figure 14-9: Liner Bond Response Graph

From the response graph, it is obvious that control factor D has the most significant effect on the bonding strength between the two types of liners. All of the other factors appear to represent the noise inherent in the manufacturing process and/or the error in the experiment. These results were validated by the manufacturer in a confirmation run.

The experiment indicated the strength of the bond was improved by establishing the hot plate temperature at 190° Fahrenheit. This setting was the upper limit of temperature allowable in the process. If heated beyond this temperature, the wax would become too fluid and be forced out from between the liners by the application roller and require frequent line shutdowns for cleanup. Even though the experimental results indicated a successful bond only 75% of the time with the test method used, this was sufficient to ensure virtually 100% conformance in the actual application. The test method chosen was much more severe than the application required. The bond was also evaluated in use to ensure it did not exceed the strength of the adhesive used to affix the safety seal to the container.

The manufacturer's final action was to apply the results of this experiment to the other manufacturing lines. As noted previously, the lines were not all identical. The primary difference was the speed at which they operated. Personnel involved with the experimental design were aware that time and temperature would probably cause a significant interaction effect in the determination of bonding strength. In order to apply the results of this experiment to other lines, they first quantified this relationship as an "energy state" based upon the line subjected to experimentation. This calculation, based on the desired hot plate temperature of 190°F and the dwell time of liner material under the hot plate is shown in the following equation.

$$Temperature \times Dwell\ Time\ =\ Energy$$

$$190°F \times 3.3\ seconds\ =\ 627\ degree\text{-}seconds$$

(14-5)

To apply the results of the experiment to a different line, which ran with a dwell time of 4.0 seconds, manufacturing personnel simply divided the energy required for a good bond by the dwell time. This provided the desired operating temperature for the slower manufacturing line as shown in the equation below.

$$Energy \: / \: Dwell \: Time \: = \: Temperature$$

$$627 \: degree\text{-}seconds \: / \: 4.0 \: seconds \: = \: 156.75° \: \approx \: 157° \: F$$

(14-6)

Sequential Experimentation

The preceding two examples appear to indicate that fractional factorial arrays will always be sufficient to solve manufacturing problems. While this is very often true, to fully optimize a process a series of experiments must be performed. When our knowledge of the relationships between inputs and outputs is low, we might choose to start with fractional factorial arrays. They can be very useful to identify which inputs have significant effects on quality characteristic outputs. As we identify significant effects, and our knowledge increases, we then use full factorials to further clarify and quantify the process relationships. This process of repeated experimentation continues until we are convinced the process is optimized, or until the potential return on our investment in the experimental process is no longer attractive.

References

1. American Supplier Institute *Taguchi Methods, Introduction to Quality Engineering Course Manual,* Version 2.1, American Supplier Institute, Dearborn, MI, 1991.

2. Barker, T. B. *Engineering Quality by Design—Interpreting the Taguchi Approach,* Marcel Dekker, New York, 1990.

3. Barker, T. B. *Quality by Experimental Design,* 2nd edition, Marcel Dekker, New York, 1994.

4. Berger, R.W (ed.) and Pyzdek, T. (ed.) *Quality Engineering Handbook*, ASQC Quality Press, Milwaukee, WI, and Marcel Dekker, New York, 1992.

5. Lochner, R.H. and Matar, J.E. *Designing for Quality—An Introduction to the Best of Taguchi and Western Methods of Statistical Experimental Design*, Quality Resources, New York and ASQC Quality Press, 1990.

6. Montgomery, D.C. and Myers, R.H. *Response Surface Methodology —Process and Product Optimization Using Designed Experiments,* John Wiley & Sons, New York, 1995.

7. Taguchi, G. *System of Experimental Design,* UNIPUB/Kraus International Publications, White Plains, NY, and American Supplier Institute, Dearborn, MI, 1987.

8. Taguchi, G. *Introduction to Quality Engineering*, Asian Productivity Organization, Tokyo, 1986.

15

Response Analysis

Thus far we have concentrated on creating and understanding experimental arrays. This is important, because the design and performance of an experiment will ultimately determine the usefulness of the results. Now we turn our attention to analysis of experimental results. There are many methods which might be employed for response analysis. While more accurate (and complex) methods are available, we shall limit our discussion in this chapter to graphical analysis techniques. In most instances, these are sufficient to facilitate process improvement efforts.

Response Tables

As illustrated in previous chapters, the first step in experimental response analysis is the calculation of average effects for control factors and, in a full factorial array, interactions. In this section, we briefly examine how we can create response tables for the calculation of average level effects. Consider the experimental array shown at the top of the following page. This illustrates a full factorial experimental array which has been run in two replications, with the average response column already calculated.

231

tc	A	B	AB	Run 1	Run 2	Average
(1)	-1	-1	+1	12	13	12.5
a	+1	-1	-1	14	14	14.0
b	-1	+1	-1	13	12	12.5
ab	+1	+1	+1	15	14	14.5

To calculate the average effect of control factor A at the -1 level, we employ the shading technique used in previous chapters:

tc	A	B	AB	Run 1	Run 2	Average
(1)	-1	-1	+1	12	13	12.5
a	+1	-1	-1	14	14	14.0
b	-1	+1	-1	13	12	12.5
ab	+1	+1	+1	15	14	14.5

Control factor A at the +1 level is shown below:

tc	A	B	AB	Run 1	Run 2	Average
(1)	-1	-1	+1	12	13	12.5
a	+1	-1	-1	14	14	14.0
b	-1	+1	-1	13	12	12.5
ab	+1	+1	+1	15	14	14.5

If we copy the treatment combination column and the response column, and place them next to the columns for A_{-1} and for A_{+1} from the preceding arrays, we obtain the following partial response table:

tc	Average	A	A
(1)	12.5	-1	-1
a	14.0	+1	+1
b	12.5	-1	-1
ab	14.5	+1	+1

Figure 15-1: Partial Response Table, Factor A

Notice that none of the averages in the response column have been shaded. Two of them correspond to the A_{-1} effect and the other two correspond to the A_{+1} effect.

For control factor B, the average effects at the -1 level and the +1 level are shown in the following two arrays:

tc	A	B	AB	Run 1	Run 2	Average
(1)	-1	-1	+1	12	13	12.5
a	+1	-1	-1	14	14	14.0
b	-1	+1	-1	13	12	12.5
ab	+1	+1	+1	15	14	14.5

tc	A	B	AB	Run 1	Run 2	Average
(1)	-1	-1	+1	12	13	12.5
a	+1	-1	-1	14	14	14.0
b	-1	+1	-1	13	12	12.5

If we now copy the B_{-1} column and the B_{+1} column from the arrays illustrated above, and add them both to our partial response table, we obtain the table shown as Figure 15-2.

tc	Average	A	A	B	B
(1)	12.5	-1	-1	-1	-1
a	14.0	+1	+1	-1	-1
b	12.5	-1	-1	+1	+1
ab	14.5	+1	+1	+1	+1

Figure 15-2: Partial Response Table, Factors A and B

The average AB interaction effects are shown below.

tc	A	B	AB	Run 1	Run 2	Average
(1)	-1	-1	+1	12	13	12.5
a	+1	-1	-1	14	14	14.0
b	-1	+1	-1	13	12	12.5
ab	+1	+1	+1	15	14	14.5

tc	A	B	AB	Run 1	Run 2	Average
(1)	-1	-1	+1	12	13	12.5
a	+1	-1	-1	14	14	14.0
b	-1	+1	-1	13	12	12.5
ab	+1	+1	+1	15	14	14.5

Adding the AB columns to the table in Figure 15-2, we obtain:

tc	Average	A	A	B	B	AB	AB
(1)	12.5	-1	-1	-1	-1	+1	+1
a	14.0	+1	+1	-1	-1	-1	-1
b	12.5	-1	-1	+1	+1	-1	-1
ab	14.5	+1	+1	+1	+1	+1	+1

Figure 15-3: Partial Response Table for A, B, and AB

Next, we move the -1 and +1 notation from the trial rows to the column headings, and add three rows labeled "Sum," "Average," and "Difference" as shown in Figure 15-4, and leave the shading in place.

Trial	Average	A_{-1}	A_{+1}	B_{-1}	B_{+1}	AB_{-1}	AB_{+1}
(1)	12.5						
a	14.0						
b	12.5						
ab	14.5						
Sum							
Ave							
Difference							

Figure 15-4: Response Table[5]

[5] Reprinted from *Designing for Quality: An Introduction to the Best of Taguchi and Western Methods of Statistical Experimental Design*, Copyright ©1990 by Robert H. Lochner and Joseph E. Matar, with permission of the publisher, Quality Resources, New York, NY.

The response table is completed by first carrying the "Averages" across each row as shown in Figure 15-5.

Trial	Average	A_{-1}	A_{+1}	B_{-1}	B_{+1}	AB_{-1}	AB_{+1}
(1)	12.5	12.5		12.5			12.5
a	14.0		14.0	14.0		14.0	
b	12.5	12.5			12.5	12.5	
ab	14.5		14.5		14.5		14.5
Sum							
Ave							
Difference							

Figure 15-5: Response Table

For each column, the sum and average are calculated. The differences between the +1 and -1 levels are then computed as illustrated below (+1 level minus -1 level).

Trial	Average	A_{-1}	A_{+1}	B_{-1}	B_{+1}	AB_{-1}	AB_{+1}
(1)	12.5	12.5		12.5			12.5
a	14.0		14.0	14.0		14.0	
b	12.5	12.5			12.5	12.5	
ab	14.5		14.5		14.5		14.5
Sum	53.5	25.0	28.5	26.5	27.0	26.5	27.0
Ave	13.375	12.5	14.25	13.25	13.5	13.25	13.5
Difference		1.75		.25		.25	

Figure 15-6: Completed Response Table

The "Difference" row allows us to see which factors have strong influence on the output characteristic measured. The "Average" row is for the preparation of response graphs. This type of response table minimizes opportunities for error. The calculations are ideal for inclusion in a computer spreadsheet, or for manual calculation on a blank form.

Response Graphs

Response graphs were constructed in previous units. Although we have only reviewed one form thus far, at least two equivalent formats exist. Consider the sample experimental responses shown below:

Average	A	B	C	D	E	F	G
Level 1	5	3	6	4	5	6	8
Level 2	7	9	6	8	7	6	4
Difference	2	6	0	4	2	0	4

The response graph for this experiment is obtained by plotting the control factor average responses for levels 1 and 2, and the overall average response as follows:

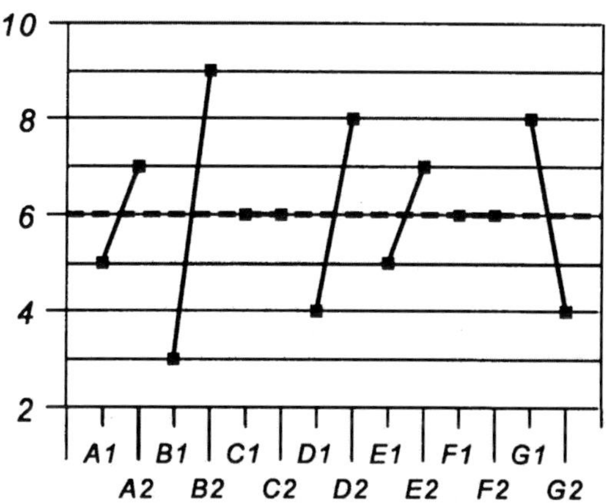

Figure 15-7: Sample Response Graph

Figure 15-7 allows us to determine both the magnitude and slope of control factor effects. The slope, however, only has value if the control factors are continuous. With this type of response graph and continuous variables, it is recommended to keep the 1 or -1 level at the low setting for the control factor and the 2 or +1 level at the high setting.

An alternate form for response graphs is shown in Figure 15-8. This form displays the magnitude of the control factor effect as a vertical line, eliminating slope. Labeling of the lines is required at both top and bottom to insure the correct level is associated with each average response point plotted on the graph. The choice of which type of response graph to use is a matter of personal preference. Both display the information necessary for process improvement and optimization efforts.

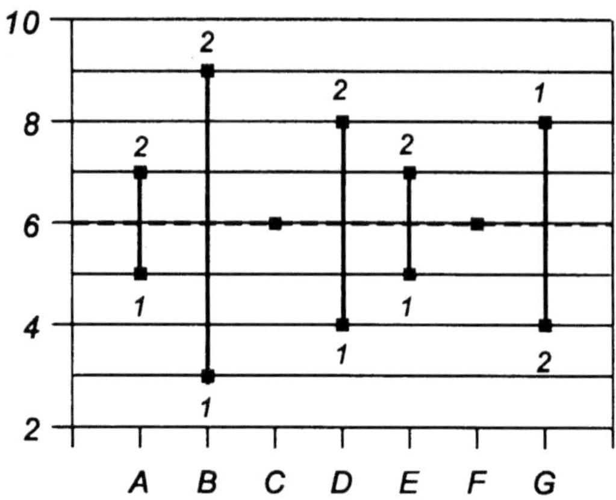

Figure 15-8: Sample Response Graph, Alternate Form

Average Response

The calculation of average response was originally presented earlier in Equation 10-5, and is repeated as 15-1.

$$\text{Average Response} = \bar{T} = \frac{\displaystyle\sum_{i=1}^{n} R_i}{n} \tag{15-1}$$

For all two-level experiments, the average line will be midway between the response values for levels 1 and 2. This line can easily be used to verify our response calculations with a quick visual scan.

Optimum Response

The response graph provides an illustration of each factor's effect on the characteristic response being measured. The next step is interpretation of the response graph to select the factors and levels needed to attain the desired response value. For instance, consider the response graph shown below for a fractional factorial experiment:

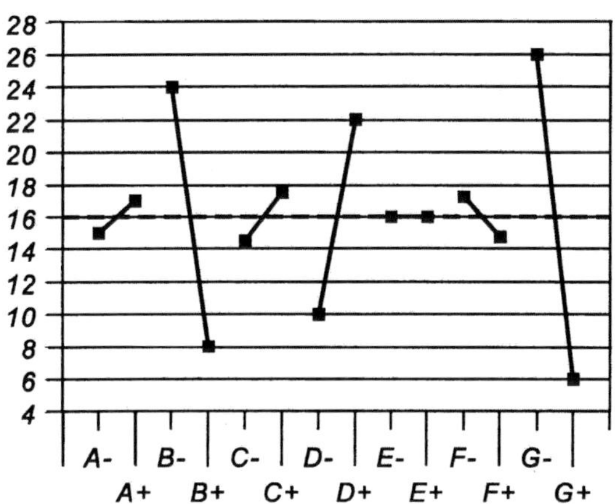

Figure 15-9: Sample Response Graph

Visually, factors B, D, and G appear to have strong effects on the characteristic response being measured. For example, when G is at the +1 level, the average response is 6—this is quite low compared to the overall average response of 16.

Factors A, C, E and F have weak or even nonexistent effects. These factors are an indication of experimental error. Weak effects do not control the process response, therefore, they are candidates for relaxed specifications and controls. On the other hand, their levels may be chosen to minimize cost.

If we desire to minimize the response represented by Figure 15-9, we choose the control factors with strong effects at levels which minimize the response. This is illustrated in Figure 15-10 by circling the desired average responses.

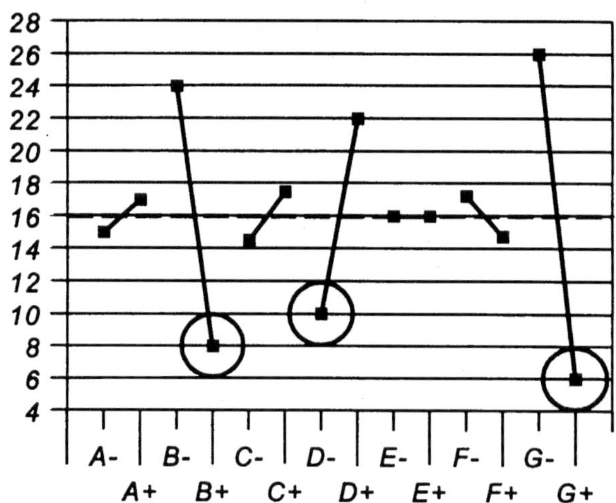

Figure 15-10: Sample Response Graph

We summarize the selection by stating the combination of factors and levels which produce the optimum response according to the experimental results.

Optimum Response: $B_+ D_- G_+$

The term "optimum response" refers only to the optimum response obtainable from the analysis of this experiment. It is generally not the absolute optimum response for the manufacturing process. Finding an absolute optimum usually requires the design and completion of multiple experiments.

If weak effects are to be set at specific levels to minimize purchase cost or power consumption or some other manufacturing parameter, they should be included in the statement of optimum response with a brief explanation. Suppose control factor A consisted of two different sources of material supply, with the +1 level being the least expensive. Our selected optimum combination would appear as follows:

Optimum Response: $A_+ B_+ D_- G_+$ (A_+ selected to minimize cost)

If all of the factors we have selected as significant have an effect on the output characteristic, we should be able to predict what the output response will be when our control factors are run at the selected levels.

Predicted Response—Main Effects

In full factorial orthogonal arrays, the analysis of any factor effect is independent of all other effects. In fractional factorials, we assume one or more interactions are not significant, and substitute additional control factors in their place. If this assumption is true, then the control factors in a fractional factorial array should also be independent. Consider the response graph shown in Figure 15-11

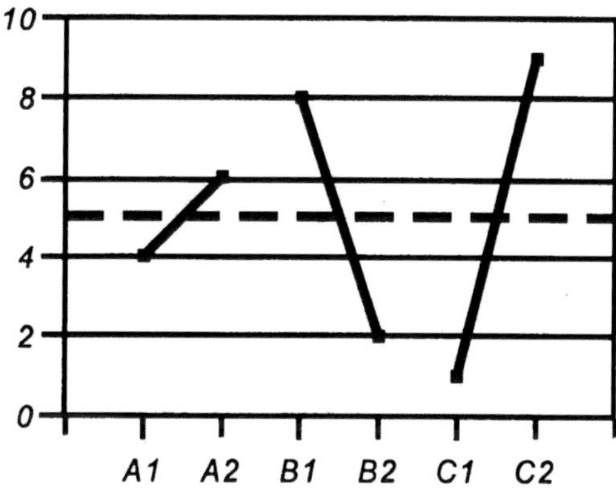

Figure 15-11: Sample Response Graph

Control factors B and C have a stronger influence on the quality characteristic output than control factor A. Factors B and C are called *strong effects*. If we desire to increase the response, we should set factor B at level 1 and factor C at level 2. The magnitude of the predicted response[6] may be calculated by using Equation 15-2, as follows:

$$\hat{\mu} = \overline{T} + (B_1 - \overline{T}) + (C_2 - \overline{T})$$

$$\text{where } \hat{\mu} = \textit{predicted response}$$
$$\overline{T} = \textit{average response} \tag{15-2}$$
$$B_1 = \textit{control factor level effect}$$
$$C_2 = \textit{control factor level effect}$$

Inserting numeric values from Figure 15-11 into Equation 15-2 yields the following predicted response:

$$\hat{\mu} = \overline{T} + (B_1 - \overline{T}) + (C_2 - \overline{T})$$
$$\hat{\mu} = 5 + (8 - 5) + (9 - 5)$$
$$\hat{\mu} = 5 + 3 + 4 \tag{15-3}$$
$$\hat{\mu} = 12$$

The response graph indicates the average effect of each individual control factor. If the interactions in this fractional factorial array are insignificant, then the control factor effects will be independent of one another and additive. For our example, setting control factor B at level 1 tends to raise the average response by three units, and setting control factor C at level 2 tends to raise the average response by four units. If we choose both levels B_1 and C_2 we expect to change the response by a total of seven units. This concept is illustrated in Figure 15-12.

---

[6] ©Copyright, American Supplier Institute, Inc., Allen Park, Michigan (U.S.A.) "Reproduced by permission under License No. 970101."

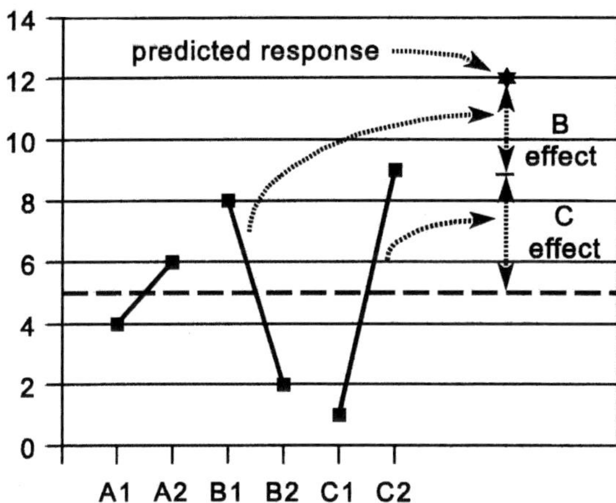

Figure 15-12: Sample Response Graph

This knowledge enables us to target the average characteristic response in a variety of ways. For example, perhaps an average response of 5 is too low, and 12 is too high. We might choose to run the process with B_2 and C_2 - this would yield a response of 6 $(6 = 5 + (2\text{-}5) + (9\text{-}5))$. If we identify a larger number of control factors that are strong effects, we increase the possible combinations of factor level settings to target a desired response. Also, if factors B and C are continuous, we may target a precise response by setting one or both factors at a level other than one of the two selected for the experiment. For example, suppose we desired an output with a value of 9. We should be able to attain this target by establishing control factor C at level 2, and establishing control factor B at a level midway between B_1 and B_2.

Predicted Response—Interactions

Equation 15-2 works fine for experiments where only main effects are analyzed. The calculation and selection of control factor levels is more complicated if interactions are present and found to have a strong influence on the quality characteristic response. For example, consider the following experimental array and response table.

tc	A	B	AB	Average
(1)	-1	-1	+1	105
a	+1	-1	-1	113
b	-1	+1	-1	96
ab	+1	+1	+1	122

Trial	Average	A_{-1}	A_{+1}	B_{-1}	B_{+1}	AB_{-1}	AB_{+1}
(1)	105	105		105			105
a	113		113	113		113	
b	96	96			96	96	
ab	122		122		122		122
Sum	436	201	235	218	218	209	227
Ave	109	100.5	117.5	109	109	104.5	113.5
Difference		17		0		9	

The response graph is completed by plotting the -1 and +1 levels from the "Ave" row in the response table, and the overall average value as shown in Figure 15-13 on the following page.

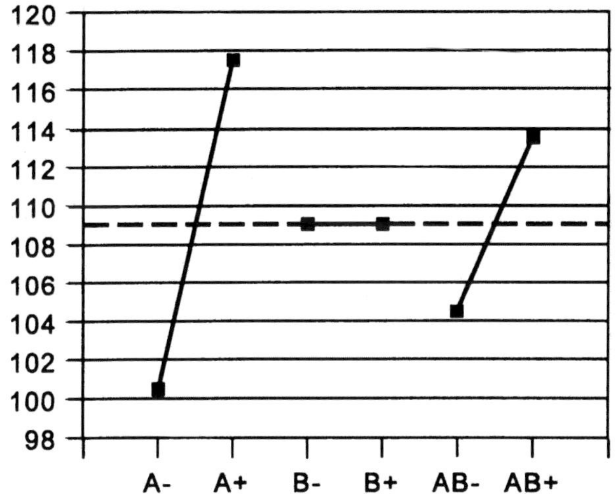

Figure 15-13: Response Graph with Interaction Effect

From the response graph, the control factor A and the interaction AB are both significant. One way to predict the maximum response is to substitute the average effect of AB at the +1 level into the same equation we used for main effects. This is illustrated in Equation 15-4.

$$\hat{\mu} = \overline{T} + (A_{+1} - \overline{T}) + (AB_{+1} - \overline{T})$$
$$\hat{\mu} = 109 + (117.5 - 109) + (113.5 - 109)$$
$$\hat{\mu} = 109 + 8.5 + 4.5 \qquad (15\text{-}4)$$
$$\hat{\mu} = 122$$

This prediction equation does predict the correct response for the combined effects of control factor A and the AB interaction both at the +1 level. There is, however, a slight problem. The prediction equation does not tell us how to achieve the AB_{+1} interaction effect.

Obtaining the predicted responses when interactions are significant requires the completion of an interaction graph. Since the response table indicated AB might be a strong effect, we examine the response for each combination of control factors in the interaction. This is found by reviewing each trial of the experimental array and its average response.

$$A_{-1}B_{-1} = 105$$
$$A_{+1}B_{-1} = 113$$
$$A_{-1}B_{+1} = 96$$
$$A_{+1}B_{+1} = 122$$

The response graph for the AB interaction effect is prepared from the information above, as illustrated in Figure 15-14. The lines cross, indicating a strong interaction.

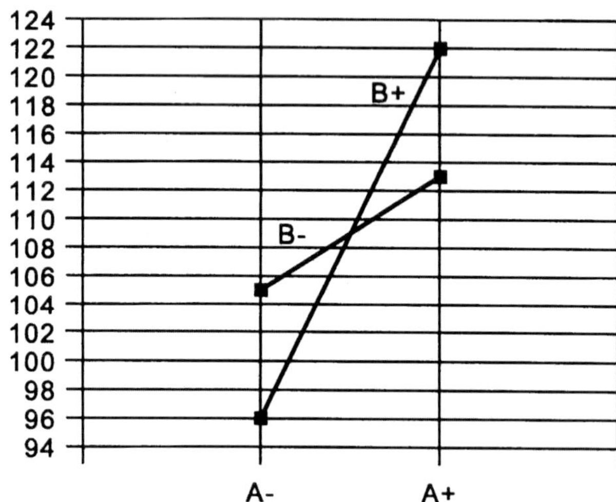

Figure 15-14: AB Interaction Graph

The $A_{+1}B_{+1}$ combination produces the maximum response with a magnitude of 122. This is the same value predicted in Equation 15-4 for the combination of the A_{+1} main effect and the AB_{+1} average interaction effect. Notice that the AB_{+1} average interaction effect is not the same as the $A_{+1}B_{+1}$ interaction response. The AB_{+1} average interaction effect is found from our experimental array by summing trials 1 and 4 together and dividing by 2. The $A_{+1}B_{+1}$ interaction response is simply the individual trial response when factor A is at the $+1$ level and factor B is at the $+1$ level. It includes the effects of both A_{+1} and B_{+1}. If we use the $A_{+1}B_{+1}$ interaction response in our prediction equation, we must subtract the A_{+1} main effect and the B_{+1} main effect from the $A_{+1}B_{+1}$ interaction response, as shown in Equation 15-5.

$$\hat{\mu} = \bar{T} + (A_{+1} - \bar{T}) + \left[(A_{+1}B_{+1} - \bar{T}) - (A_{+1} - \bar{T}) - (B_{+1} - \bar{T}) \right]$$

$$\hat{\mu} = \bar{T} + A_{+1} - \bar{T} + A_{+1}B_{+1} - \bar{T} - A_{+1} + \bar{T} - B_{+1} + \bar{T}$$

$$\hat{\mu} = A_{+1}B_{+1} - B_{+1} + \bar{T} \tag{15-5}$$

$$\hat{\mu} = 122 - 109 + 109$$

$$\hat{\mu} = 122$$

This is the same prediction we obtained when we used the A_{+1} and the AB_{+1} average interaction. The difference is that (15-5) clearly indicates which levels of both A and B are needed to obtain the desired response. Of course, completing the AB interaction graph shown in Figure 15-14 provides the needed information. Our optimum response will be stated as follows:

$$\text{Optimum Response: } A_{+1} \ (B_{+1}) \ AB_{+1}$$

Factor B has been placed in parentheses to indicate it is required because of its role in the interaction, even though B itself is not a strong effect. Including the AB_{+1} interaction indicates it is significant and that an interaction graph was necessary to determine the levels of A and B.

Equation 15-4 is an example of prediction equations with interactions—it may appear as if we could have merely selected the experimental trial with the highest response. The array we used to demonstrate predictions was a full factorial. There was no need to predict responses because all possible control factor combinations had already been run. The response column contains the actual value which resulted from each control factor combination. (We need only to review the actual responses, and select the trial with the response closest to output value we desire.) With fractional factorials, and especially with large fractional factorial arrays, the solution is not always obvious. In these cases, the prediction equation must be used.

Confirmation Runs

To verify the validity of experimental results and predictions, it is necessary to perform at least one additional trial after completing the response graphs and prediction equations. This trial is a test of the accuracy of the predictions. It confirms or refutes the results of the experimental analysis. This trial is known as the *confirmation run*[7].

To complete a confirmation run, simply select the optimum factor levels for all strong effects, calculate the predicted response for the characteristic being evaluated, and run one additional trial with the control factors set at the selected levels. If the actual response agrees with the predicted response, the validity of the experiment is confirmed. (Methods exist to estimate experimental error and assign confidence intervals for predicted responses, however, they are beyond the scope of this text.)

If the actual response does not agree with the predicted response, some factors or interactions assumed significant were actually not (or vice versa). You should re-evaluate the prediction calculation excluding some of the factors you assumed were significant or including some of the factors you did not select previously. If the actual response still does not agree with the predicted response there may be other factors (control factors, noise factors, or interactions) affecting the quality characteristic response that have not been included in the experiment.

------

[7] ©Copyright, American Supplier Institute, Inc., Allen Park, Michigan (U.S.A.) "Reproduced by permission under License No. 970101."

Multiple Response Analysis

Thus far, we have discussed experimental arrays and response analysis in terms of a single quality characteristic response. It is quite normal for experiments to be designed and performed with the goal of improving a single characteristic, however, this represents only a small portion of the return possible from these investigations. Whenever an experiment is performed, it is an opportunity to quantify the relationships between control factors and multiple characteristic responses. Knowledge can be obtained about more than one response concurrently, with the only additional cost being the cost of measuring and recording additional characteristic values. For example, consider the following array which shows only length as a quality characteristic response (response columns for each replication are not shown):

tc	A	B	AB	Average Length
(1)	-1	-1	+1	45
a	+1	-1	-1	40
b	-1	+1	-1	47
ab	+1	+1	+1	43

Figure 15-15: Array with Single Response

This array will show us which control factors actually determine the length of the part being studied, however, we risk changing other part dimensional characteristics if we treat length as the sole output quality characteristic. Also, if we choose not to measure characteristics such as width and height, we are missing an opportunity to quantify the input/output relationship for these other characteristics with minimal additional expenditure.

We record multiple responses by simply adding additional columns to our array, as illustrated in Figure 15-16. (Replicates not shown.)

tc	A	B	AB	Average Length	Average Width	Average Height
(1)	-1	-1	+1	45	20	33
a	+1	-1	-1	40	17	34
b	-1	+1	-1	47	22	35
ab	+1	+1	+1	43	18	34

Figure 15-16: Array with Multiple Responses

To fully analyze this array, we must produce response tables and response graphs for each quality characteristic output separately. For example, the response tables for length, width, and height are shown in Figures 15-17, 15-18, and 15-19, respectively.

Trial	Ave	A_{-1}	A_{+1}	B_{-1}	B_{+1}	AB_{-1}	AB_{+1}
(1)	45	45		45			45
a	40		40	40		40	
b	47	47			47	47	
ab	43		43		43		43
Sum	175	92	83	85	90	87	88
Ave	43.75	46	41.5	42.5	45	43.5	44
Difference		-4.5		2.5		0.5	

Figure 15-17: Response Table for Length

Trial	Ave	A_{-1}	A_{+1}	B_{-1}	B_{+1}	AB_{-1}	AB_{+1}
(1)	20	20		20			20
a	17		17	17		17	
b	22	22			22	22	
ab	18		18		18		18
Sum	77	42	35	37	40	39	38
Ave	19.3	21	17.5	18.5	20	19.5	19
Difference		-3.5		1.5		-0.5	

Figure 15-18: Response Table for Width

Trial	Ave	A_{-1}	A_{+1}	B_{-1}	B_{+1}	AB_{-1}	AB_{+1}
(1)	33	33		33			33
a	34		34	34		34	
b	35	35			35	35	
ab	34		34		34		34
Sum	136	68	68	67	69	69	67
Ave	34	34	34	33.5	34.5	34.5	33.5
Difference		0		1.0		-1.0	

Figure 15-19: Response Table for Height

The response graph for length indicates this characteristic is controlled by the selection of levels for control factor A, and perhaps by the selection of levels for control factor B.

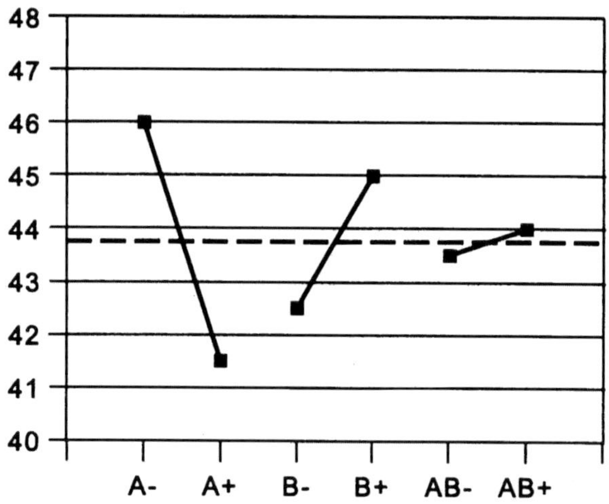

Figure 15-20: Response Graph for Length

If the target value for part length is 48, our present action would be to select factor levels to increase the length of the part. Within the range of this experiment, we would state the optimum response as follows:

Length Optimum Response: A_{-1} B_{+1}

Since the array is a full factorial design, we can assess the result of this combination by scanning the array and observing the result when A is at the -1 level and B is at the +1 level. In this case, the result is 47.

The response graph for width shown in Figure 15-21 also indicates this characteristic is controlled by factor A. If our goal is to target a value of 21 for this part, our optimum response would be:

Width Optimum Response: A_{-1}

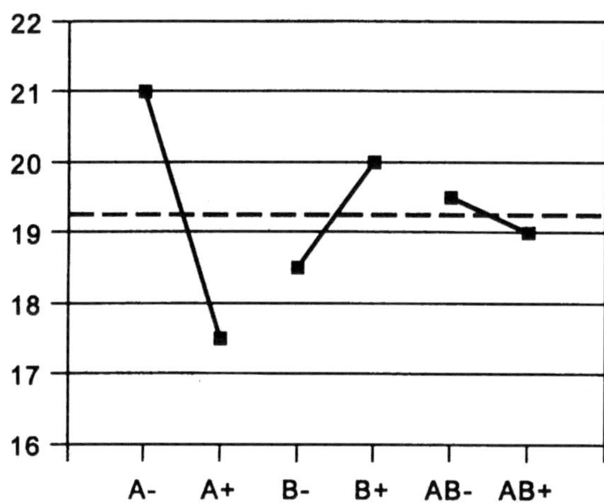

Figure 15-21: Response Graph for Width

The response graph for height, shown in Figure 15-22, indicates this characteristic is controlled differently than the other two responses. The height response is controlled by the AB interaction (and hence factors A and B). Control factor A is insignificant in affecting the height of the part by itself. The interaction graph for height is shown in Figure 15-23. As was the case with the other two characteristic responses, assume that height is also desired as large as possible.

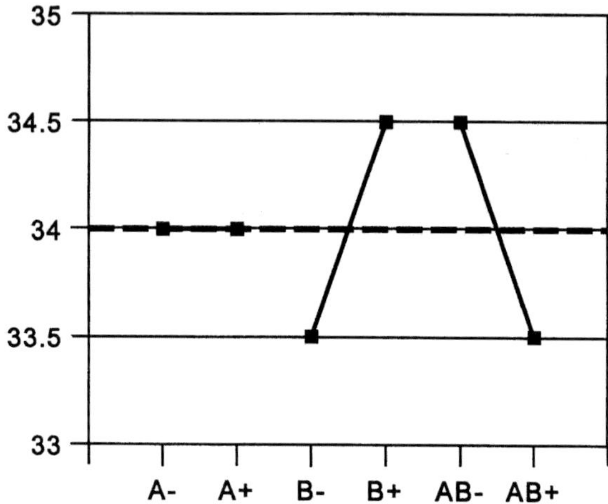

Figure 15-22: Response Graph for Height

The AB interaction graph indicates the maximum height response is obtained when A is set at the "-" level and B is set at the "+" level.

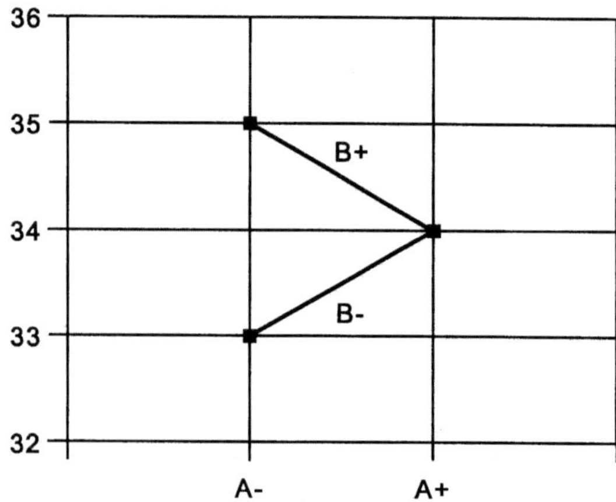

Figure 15-23: Interaction Response Graph

The optimum response for height is:

$$\textit{Height optimum response: } (A_{-1}) \; B_{+1} \; AB_{-1}$$

A_{-1} is enclosed in parentheses to indicate it was chosen solely due to the effect it has in the AB interaction.

When dealing with multiple response analysis, the last step is to tabulate all of the individual optimum responses. Figure 15-24 lists the optimum responses for the example just completed.

Response	Goal	A	B	AB
Length	→ 48 ←	A_{-1}	B_{+1}	
Width	→ 21 ←	A_{-1}		
Height	↑	(A_{-1})	B_{+1}	AB_{-1}
Overall Optimum Response		A_{-1}	B_{+1}	AB_{-1}

Figure 15-24: Multiple Response Table

The arrows in the "goal" column of the table above indicate the desired target for each quality characteristic response, in this example length and width at specified targets and height as "Larger-the-better." The row labeled "Overall Optimum Response" represents the control factor levels chosen to optimize as many of the response characteristics as possible concurrently, within the range of this experiment. For the example illustrated, it is easy to choose an overall optimum response since none of the single characteristic optimum responses conflicted with any of the others.

Quite often it is not possible to optimize all quality characteristic responses concurrently. This usually indicates the specifications which dictate nominal values for the process were conceived without adequate process knowledge, that is, the relationships between control factors and characteristic responses were not known or quantified prior to the creation of the specifications. In such a situation either the specifications should be changed, or some responses will have to be chosen for optimization at the expense of others.

Example 15.1: Multiple Response Characteristics

In Example 13.1 we examined a full factorial experiment designed to investigate control factors for bearing outside diameter. For convenience, the illustration of the split bearing is repeated below.

Figure 15-25: Bearing Configuration

In the course of performing this experiment, the manufacturer actually evaluated multiple quality characteristic response variables for concurrent optimization. In addition to outside diameter, they also examined the inside diameter response and the resulting overall bearing length. These are the three major characteristics which determine success in a customer's application. The experiment was conducted with a Taguchi L_8 array. This array is shown in Figure 15-26 with the resulting three average response columns.

T R I A L	Control Factors							Results (inches)		
	A Blank Length	B ID Tool	AB	C OD Tool	AC	BC	ABC	OD 1=.0001	ID 1=.0001	Length 1=1.0000
1	short	small	1	small	1	1	1	-10.2	10.8	.5894
2	short	small	1	large	2	2	2	-8.5	9.5	.5881
3	short	large	2	small	1	2	2	-11.5	15.2	.5910
4	short	large	2	large	2	1	1	-9.4	13.1	.5894
5	long	small	2	small	2	1	2	-9.6	9.7	.5910
6	long	small	2	large	1	2	1	-7.9	8.8	.5900
7	long	large	1	small	2	2	1	-10.9	13.9	.5929
8	long	large	1	large	1	1	2	-8.9	12.8	.5920

Figure 15-26: L_8 Experimental Array with Multiple Responses

We have already discussed bearing OD measurement in Chapter 13. Measurement of bearing ID is accomplished as a deviation above a gage setting master. This explains the magnitude of the inside diameter measurements recorded in the response table.

The control factor and interaction average responses for each quality characteristic are calculated through the use of the shaded response table for the Taguchi L_8 array. (This table is found in the Appendix.) The response tables are shown on the following three pages as Figures 15-27, 15-28, and 15-29.

		A1	A2	B1	B2	AB1	AB2	C1	C2	AC1	AC2	BC1	BC2	ABC1	ABC2
BEARING OUTSIDE DIAMETER															
Trial	Ave.	A1	A2	B1	B2	AB1	AB2	C1	C2	AC1	AC2	BC1	BC2	ABC1	ABC2
1	-10.2	-10.2		-10.2		-10.2		-10.2		-10.2		-10.2		-10.2	
2	-8.5	-8.5		-8.5		-8.5			-8.5		-8.5		-8.5		-8.5
3	-11.5	-11.5			-11.5		-11.5	-11.5		-11.5			-11.5		-11.5
4	-9.4	-9.4			-9.4		-9.4		-9.4		-9.4	-9.4		-9.4	
5	-9.6		-9.6	-9.6			-9.6	-9.6			-9.6	-9.6			-9.6
6	-7.9		-7.9	-7.9			-7.9		-7.9	-7.9			-7.9	-7.9	
7	-10.9		-10.9		-10.9	-10.9		-10.9			-10.9		-10.9	-10.9	
8	-8.9		-8.9		-8.9	-8.9			-8.9	-8.9		-8.9			-8.9
Sum		-39.6	-37.3	-36.2	-40.7	-38.5	-38.4	-42.2	-34.7	-38.5	-38.4	-38.1	-38.8	-38.4	-38.5
Average		-9.900	-9.325	-9.050	-10.17	-9.625	-9.600	-10.55	-8.675	-9.625	-9.600	-9.525	-9.700	-9.600	-9.625
Difference		-0.575		1.125		-0.025		-1.875		-0.025		0.175		0.025	
Overall Average Response															-9.6125

Figure 15-27: OD Response Table

BEARING INSIDE DIAMETER															
Trial	Ave.	A1	A2	B1	B2	AB1	AB2	C1	C2	AC1	AC2	BC1	BC2	ABC1	ABC2
1	10.8	10.8		10.8		10.8		10.8		10.8		10.8		10.8	
2	9.5	9.5		9.5		9.5			9.5		9.5		9.5		9.5
3	15.2	15.2			15.2		15.2	15.2		15.2			15.2		15.2
4	13.1	13.1			13.1		13.1		13.1		13.1	13.1		13.1	
5	9.7		9.7	9.7			9.7	9.7			9.7	9.7			9.7
6	8.8		8.8	8.8			8.8		8.8	8.8			8.8	8.8	
7	13.9		13.9		13.9	13.9		13.9			13.9		13.9	13.9	
8	12.8		12.8		12.8	12.8			12.8	12.8		12.8			12.8
Sum		48.6	45.2	38.8	55.0	47.0	46.8	49.6	44.2	47.6	46.2	46.4	47.4	46.6	47.2
Average		12.15	11.30	9.70	13.75	11.75	11.70	12.40	11.05	11.90	11.55	11.60	11.85	11.65	11.80
Difference		0.85		-4.05		0.05		1.35		0.35		-0.25		-0.15	
									Overall Average Response					11.725	

Figure 15-28: ID Response Table

BEARING LENGTH															
Trial	Ave.	A1	A2	B1	B2	AB1	AB2	C1	C2	AC1	AC2	BC1	BC2	ABC1	ABC2
1	.5894	.5894		.5894		.5894		.5894		.5894		.5894		.5894	
2	.5881	.5881		.5881		.5881			.5881		.5881		.5881		.5881
3	.5910	.5910			.5910		.5910	.5910		.5910			.5910		.5910
4	.5894	.5894			.5894		.5894		.5894		.5894	.5894		.5894	
5	.5910		.5910	.5910			.5910	.5910			.5910	.5910			.5910
6	.5900		.5900	.5900			.5900		.5900	.5900			.5900	.5900	
7	.5929		.5929		.5929	.5929		.5929			.5929		.5929	.5929	
8	.5920		.5920		.5920	.5920			.5920	.5920		.5920			.5920
Sum		2.3579	2.3659	2.3585	2.3653	2.3624	2.3614	2.3643	2.3595	2.3624	2.3614	2.3618	2.3620	2.3617	2.3621
Average		0.5895	0.5915	0.5896	0.5913	0.5906	0.5904	0.5911	0.5899	0.5906	0.5904	0.5905	0.5905	0.5904	0.5905
Difference		-0.002		-0.0017		0.00025		0.0012		0.00025		-0.00005		-0.0001	

	Overall Average Response	.59048

Figure 15-29: Length Response Table

The response graph for outside diameter is shown in Figure 15-30. It is prepared from the "average" and "overall average response" rows in the outside diameter response table.

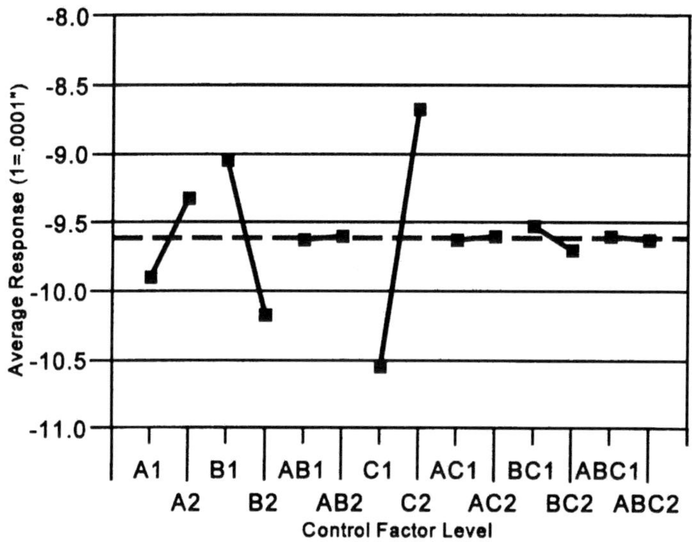

Figure 15-30: Outside Diameter Response

This response graph was reviewed briefly in Example 13-1, however, we were not prepared to discuss multiple responses at that time. The engineering specification for this quality characteristic output is 0 to -24 (in units of 0.0001"). Since the interactions in this response graph all appear insignificant, our goal is to select significant control factor levels to attain an average response of -0.0012 inches. We observe that control factor C, the outside diameter tool, appears to have the largest affect on the outside diameter. Factor B, the inside diameter tool, also appears significant. The control factor for cut blank length, factor A, is a bit more difficult to assess. We had discussed in Chapter 13 that the effect of this factor represented only 2.4% of the engineering specification. Certainly, this is small compared to the range of acceptability. However, if the manufacturing process is producing parts with exceptionally high C_p's or C_{pk}'s the width of the engineering specification may be irrelevant in determining a factor's significance. And, again, while statistical calculations do exist which may help us ascertain significance, they are somewhat complex and beyond the scope of this text. Our goal is to determine significant factors based upon a visual scan of the response graph and our process knowledge and experience.

How can we determine visually which factors are significant? As we already know, the insignificant factors are characterized by responses on the graph which are smaller in magnitude. These smaller magnitude responses are indicative of experimental error. (This term encompasses all sources such as measurement error, set-up error, or operator error.) To sort the two types of responses mentally, it sometimes helps to visualize an "error band" on the response graph. This is illustrated in Figure 15-31.

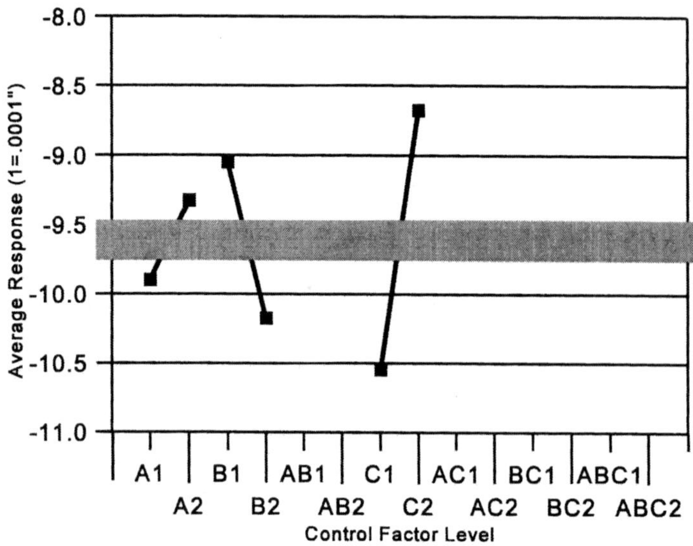

Figure 15-31: Visualizing Error

If our "error band" is correct, there are three significant control factors in this response graph. If this were a fractional factorial array, this assumption could be tested through the use of the prediction equations and confirmation runs as discussed earlier in this chapter.

Since this is a full factorial array, all possible combinations of the control factors have already been evaluated. The OD response has a target of -0.0012 inches. To ascertain the combination of control factors which produce the response closest to that value, we can review the experimental array and the response column. From the array, we obtained a value of -0.00115 inches when A was at level 1, B was at level 2, and C was at level 1. If this response is adequate, we now know at what levels to establish our control factors to attain this response.

With this particular characteristic, it was very easy to select the appropriate factor levels. Many times it is much more difficult to come close to the desired target. It is then necessary to select factors and levels in a combination which tends the move the characteristic response toward the target while simultaneously minimizing adverse effects on other measured responses. In these situations, it is necessary to have an understanding of the relationships presented on the response graphs. For instance, from the graph under evaluation, relatively large changes in the OD response can be obtained by altering factor C, the OD tool in increments equal to the one used in this experiment. If smaller adjustments to the outside diameter are desired, we might choose to increment the ID tool, or perhaps the blank length. If the response obtained in trial 3 of -0.00115 inches is not close enough, we might choose to decrease the blank length one additional increment to obtain a response closer to our target value of -0.0012 inches.

For this response, the primary control factor for targeting is the OD forming tool, the secondary control factor is the ID forming tool, and the tertiary control factor is the blank length. This assumes the increments of any changes in the control factor levels are going to be equal to the increments used in the experiment. As we shall see, the inside diameter and the overall bearing length exhibit different responses with respect to the relative importance of control factors.

The response graph for inside diameter is shown in Figure 15-31.

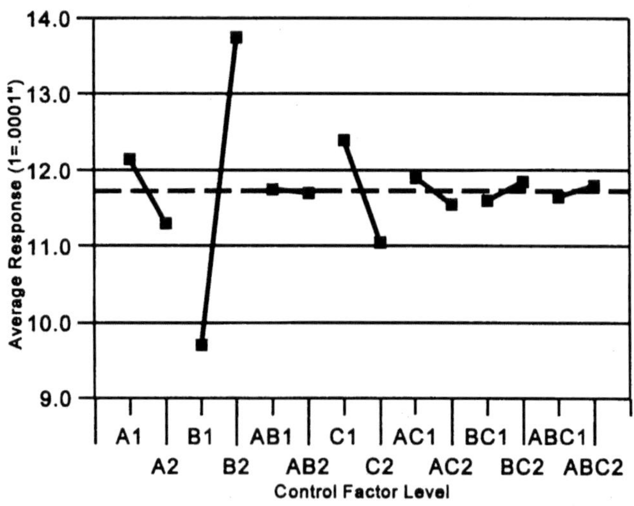

Figure 15-31: Inside Diameter Response

The target value for inside diameter is 0.0015 inches above the diameter of the gage setting master. From the responses in the experimental array, trial 3 produced the result closest to our desired target. This is the same trial which produced the response closest to the desired OD target. For the ID, it is obvious that control factor B, the ID forming tool is the primary control factor to adjust to effect changes in the ID. It appears that the OD tool may also have an effect, although it is smaller in magnitude. The blank length, control factor A, seems to be insignificant as do the interaction effects.

The response graph for overall length is shown in Figure 15-32.

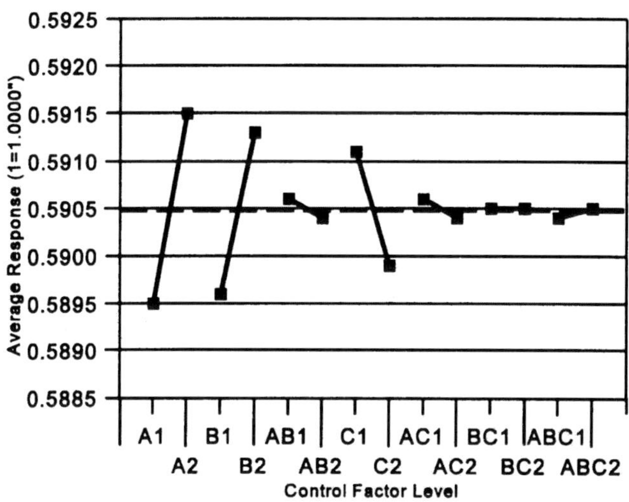

Figure 15-32: Overall Length Response

The part length specification is $0.590 \pm .010$ inches. For this quality characteristic response, all three control factors are significant. From the experimental array, the target response is achieved when these control factors are established at the levels used for trial number 6. The primary control factor is A, the blank length. The secondary control factor is B, the ID forming tool. The tertiary control factor is C, the OD forming tool. Even though this may sound ideal, the three measured responses have different primary control factors and we shall see that all three targets can not be obtained simultaneously within the range of this experiment.

Recall that the OD response required the levels established for trial number 3, namely A_1, B_2, and C_1. The ID response required the same levels of B and C. If we assume that OD and ID are the primary quality characteristics, then the factor levels selected for them will govern our overall selection.

Even though we have already predetermined the levels of the control factors, a multiple response table is shown below. This table has a dual purpose. It can be used either as a tool for selecting control factor levels or as a summary of levels selected. For complex experiments with significant interaction effects and/or many response characteristics, the multiple response table is often the only way to select factor levels in a logical manner.

Response	Goal	A	B	AB	C	AC	BC	ABC
OD	→ -.0012 ←	A_1	B_2	-	C_1	-	-	-
ID	→ .0015 ←	-	B_2	-	C_1	-	-	-
Length	→ .590 ←	A_2	B_1	-	C_2	-	-	-
Overall Optimum Response		A_1	B_2	-	C_1	-	-	-

Figure 15-33: Multiple Response Table

With the control factor levels indicated in the last row of the multiple response table, the OD and ID will both be close to their target values. The length of the part will average approximately 0.5910 inches. This is the average response when the control factors are established at $A_1B_2C_1$, as indicated in the response column of the experimental array. Since the engineering specification for part length allows a range of 0.020 inches, this average may be acceptable for this part.

Example 15.2: Predicting Coil Profile Average Response

In Example 14.2, we examined a simple experimental array to minimize the distortion of processed steel coil. We were told that the predicted response for the control factors levels established by the experiment was 0.00024 inches. We also were informed the actual average response for production coils several months later was 0.00025 inches. The L_4 array used for the coil profile experiment is repeated below as Figure 15-34.

Trial	Flattener Gap	Line Speed	Steel Supplier	Average Response
1	low	slow	1	.00035
2	low	fast	2	.00051
3	high	slow	2	.00037
4	high	fast	1	.00025

Figure 15-34: Coil Profile Array

The response table for this array and the prediction equation for the average crossbow response are both shown on the following page.

We witnessed in Chapter 13 that factors A and C were significant. This can also be determined from the relative magnitudes of the results in the "Difference" row of the response table. The predicted response for average crossbow is shown in Equation 15-6.

Trial	Ave	Gap		Speed		Supplier	
		low	high	slow	fast	1	2
1	.00035	.00035		.00035		.00035	
2	.00051	.00051			.00051		.00051
3	.00037		.00037	.00037			.00037
4	.00025		.00025		.00025	.00025	
Sum	.00148	.00086	.00062	.00072	.00076	.00060	.00088
Ave	.00037	.00043	.00031	.00036	.00038	.00030	.00044
Difference		-.00012		.00002		.00014	

Figure 15-35: Coil Profile Response Table

$$\hat{\mu} = \overline{T} + (A_2 - \overline{T}) + (C_1 - \overline{T})$$

$$\hat{\mu} = .00037 + (.00031 - .00037) + (.00030 - .00037)$$

$$\hat{\mu} = .00037 - .00006 - .00007$$

$$\hat{\mu} = .00024 \ inches$$

(15-6)

References

1. American Supplier Institute *Taguchi Methods, Introduction to Quality Engineering Course Manual,* Version 2.1, American Supplier Institute, Dearborn, MI, 1991.

2. Barker, T. B. *Engineering Quality by Design—Interpreting the Taguchi Approach*, Marcel Dekker, New York, 1990.

3. Barker, T. B. *Quality by Experimental Design,* 2nd edition, Marcel Dekker, New York, 1994.

4. Berger, R.W (ed.) and Pyzdek, T. (ed.) *Quality Engineering Handbook*, ASQC Quality Press, Milwaukee, WI, and Marcel Dekker, New York, 1992.

5. Lochner, R.H. and Matar, J.E. *Designing for Quality—An Introduction to the Best of Taguchi and Western Methods of Statistical Experimental Design*, Quality Resources, New York and ASQC Quality Press, 1990.

6. Montgomery, D.C. and Myers, R.H. *Response Surface Methodology —Process and Product Optimization Using Designed Experiments,* John Wiley & Sons, New York, 1995.

7. Taguchi, G. *System of Experimental Design,* UNIPUB/Kraus International Publications, White Plains, NY, and American Supplier Institute, Dearborn, MI, 1987.

8. Taguchi, G. *Introduction to Quality Engineering*, Asian Productivity Organization, Tokyo, 1986.

16

Minimizing Variation

We have examined how to design and analyze experiments to target our outputs. This is often the goal of process improvement efforts for both smaller-the-better and larger-the-better characteristics. For many processes, manufacturing personnel have a good understanding of how to attain a specific target value. Minimizing variation in the output, however, is usually less obvious. In this chapter, we shall discuss how to minimize variation in our process output characteristics. We shall also examine how to make our process resistant to noise factors. Once we accomplish this, the outputs of the process will maintain their target values and exhibit minimal variation even when subjected to noise factors. This type of process is called "robust."

Signal-to-Noise Ratios

To minimize variation, we shall employ Taguchi's signal-to-noise ratio calculations[8] in our analysis of experimental results. Before we begin, a

---

[8] ©Copyright, American Supplier Institute, Inc., Allen Park, Michigan (U.S.A.) "Reproduced by permission under License No. 970101."

brief explanation of this concept is required. The signal-to-noise ratio, or SNR, is a measure of the consistency of the output response. This ratio is used in our response tables and graphs in the same fashion as the average level effects we have used for targeting. For a manufacturing process, the concepts of signal and noise might be illustrated by the following figure. As the signal becomes larger, or the noise becomes smaller, the SNR becomes larger. This indicates improved consistency in response characteristics.

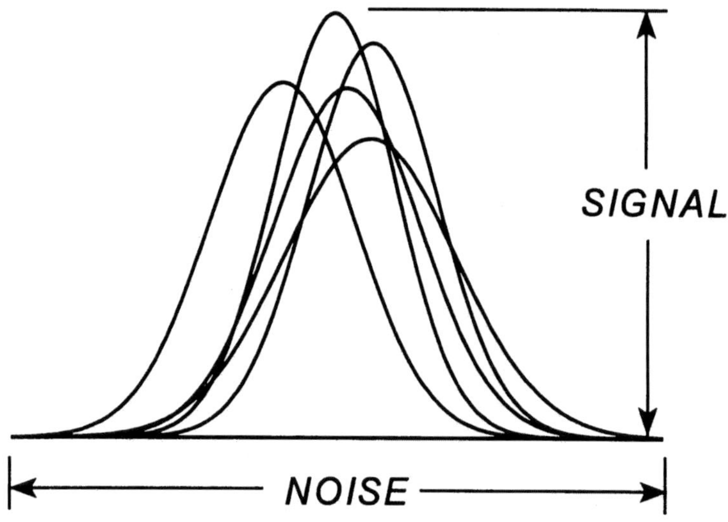

Figure 16-1: Signal-to-Noise Concept

The illustration above represents a process which is not in statistical control. We can increase the signal by reducing the variation of the normal curves illustrated above. This will cause the heights of the curves to increase. We can decrease the noise either by reducing the variation, or by removing the assignable causes which precipitate inconsistent targeting. Either of these actions will increase the SNR of the process output. As we shall examine in this chapter, when we use SNR's to analyze experimental responses, we choose control factor or interaction levels which produce the highest signal-to-noise ratios. The SNR simply tells us which levels to choose to improve our process.

The equations for signal-to-noise ratio do not discern whether specific factor or interaction levels decrease standard deviation or if they remove assignable cause variation. The SNR equations merely indicate which experimental settings produce the most consistent responses. We shall review some experimental results which demonstrate the removal of variation and the resulting improvements in signal-to-noise ratio.

The signal-to-noise ratio equations are calculated with logarithms. The unit of measure for all of these equations is decibels (dB).

Larger-the-Better SNR

The calculation of signal-to-noise ratio for a larger-the-better output characteristic is shown in Equation 16-1. While the calculation can become somewhat lengthy for large amounts of data, it is easily programmed into a calculator or personal computer.

$$SNR\ (dB) = -10 \log \left(\frac{\sum_{i=1}^{n} \frac{1}{y_i^2}}{n} \right)$$

$$(16\text{-}1)$$

$$\text{where } y_i = \textit{individual measured output}$$
$$n = \textit{number of measured outputs}$$

The calculation of SNR for one trial of a designed experiment with four replicates is illustrated in Equation 16-2.

Measured Outputs: 15, 19, 17, 14

$$SNR\,(dB) = -10\log\left(\frac{\sum\limits_{i=1}^{n}\dfrac{1}{y_i^2}}{n}\right)$$

$$SNR\,(dB) = -10\log\left(\frac{\dfrac{1}{15^2}+\dfrac{1}{19^2}+\dfrac{1}{17^2}+\dfrac{1}{14^2}}{4}\right)$$

$$SNR\,(dB) = -10\log\left(\frac{\dfrac{1}{225}+\dfrac{1}{361}+\dfrac{1}{289}+\dfrac{1}{196}}{4}\right)$$

$$(16\text{-}2)$$

$$SNR\,(dB) = -10\log\,(3.944 \times 10^{-3})$$

$$SNR\,(dB) = 24.04$$

Smaller-the-Better SNR

The calculation of signal-to-noise ratio for a smaller-the-better output characteristic is shown in Equation 16-3.

$$SNR\,(dB) = -10\log\left(\frac{\sum\limits_{i=1}^{n}y_i^2}{n}\right)$$

$$(16\text{-}3)$$

$$\text{where } y_i = \textit{individual measured output}$$

$$n = \textit{number of measured outputs}$$

For illustrative purposes, using the same group of numbers we did for the previous example, the signal-to-noise ratio calculation for a smaller-the-better characteristic is shown in Equation 16-4.

Measured Outputs: 15, 19, 17, 14

$$SNR\,(dB) = -10\log\left(\frac{\sum\limits_{i=1}^{n} y_i^2}{n}\right)$$

$$SNR\,(dB) = -10\log\left(\frac{15^2 + 19^2 + 17^2 + 14^2}{4}\right)$$

$$SNR\,(dB) = -10\log\left(\frac{225 + 361 + 289 + 196}{4}\right)$$

(16-4)

$$SNR\,(dB) = -10\log\,(267.75)$$

$$SNR\,(dB) = -24.28$$

The sample calculations illustrate an important characteristic of these two signal-to-noise ratios. Both were calculated with the same group of data, although we defined one set of outputs as a larger-the-better and one as a smaller-the-better characteristic. We obtain two different results because of the structure of Taguchi's equations.

If we substitute larger numbers with the same range into these two equations, we would find the signal-to-noise ratio for the larger-the-better characteristic increases and the ratio decreases for the smaller-the-better characteristic. Conversely, if we substitute smaller numbers into the equations, while maintaining the same range, the signal-to-noise ratio for the larger-the-better characteristic becomes smaller and the ratio for the smaller-the-better characteristic becomes larger. If we substitute data sets with smaller range and the same average into both equations, we would find both equations produce larger signal-to-noise ratios.

For these SNR's the desired target of the process output is not disregarded as it would be in a calculation of standard deviation. The goal of process optimization is not only the minimization of standard deviation. We desire to minimize variation about the desired target value.

Nominal-the-Best SNR

For this type of output characteristic, we may find some control factors which affect targeting, and some which affect variation. It is important to perform our calculations in a method which allows us to discern the difference. For optimizing manufacturing processes, we shall identify factors which affect targeting in the normal fashion, that is, through response graphs as illustrated in previous chapters. For some applications, such as Research & Development, readers may want to consult Dr. Taguchi's works listed at the end of this chapter. In R&D environments, it may be wise to identify targeting control factors through the use of the "sensitivity analysis" approach presented by Taguchi.

Two equations exist for the signal-to-noise ratio of a nominal-the-best characteristic. The proper equation to use is determined by the possible measured values of the output response. One equation (Type I) is for outputs whose values are always positive, and another (Type II) is for outputs which can assume either negative or positive values.

Nominal-the-Best SNR—Type I

For this category of output characteristic, where the measured values are always positive, the signal-to-noise ratio calculation is shown below in Equation 16-5.

$$SNR\ (dB) = 10\log\left(\frac{\bar{y}^2}{\sigma^2} - \frac{1}{n}\right)$$

(16-5)

$$where\ \bar{y} = mean\ of\ measured\ output$$

$$\sigma^2 = sample\ variance$$

$$n = number\ of\ measured\ outputs$$

To illustrate this SNR calculation, we shall once again use the four data points introduced earlier. This time, we assume the data represents the response of a positive-only nominal-the-best output. The calculation is shown in Equation 16-6.

Measured outputs: 15, 19, 17, 14

$$\bar{y} = \frac{15 + 19 + 17 + 14}{4} = 16.25$$

$$\sigma^2 = \frac{\sum_{i=1}^{n} (y_i - \bar{y})^2}{(n-1)}$$

$$\sigma^2 = \frac{(15-16.25)^2 + (19-16.25)^2 + (17-16.25)^2 + (14-16.25)^2}{(4-1)}$$

$$\sigma^2 = 4.92$$

$$SNR\ (dB) = 10 \log\left(\frac{\bar{y}^2}{\sigma^2} - \frac{1}{n} \right)$$

$$SNR\ (dB) = 10 \log\left(\frac{(16.25)^2}{(4.92)} - \frac{1}{4} \right)$$

$$SNR\ (dB) = 10 \log (53.71 - .25)$$

$$S/N\ (dB) = 17.28$$

(16-6)

Nominal-the-Best SNR—Type II

For this category of nominal-the-best output, where the data values can be either positive or negative, and the signal-to-noise calculation is based upon the sample variance. This is shown in Equation 16-7.

$$SNR\ (dB) = -10\ \log \left(\frac{\sum\limits_{i=1}^{n} (y_i - \bar{y})^2}{n-1} \right) = -10\ \log\ \sigma^2$$

$$(16\text{-}7)$$

$$where\ y_i = individual\ measured\ output$$
$$\bar{y} = mean\ of\ measured\ output$$
$$n = number\ of\ measured\ outputs$$

Example 16.1: Bearing Characteristics

Example 13.1, for a split bearing, was actually performed with varying levels of input material thickness. The goals of the experiment were:

1) quantify the effect of each tool on characteristic responses for various input material thicknesses
2) determine the optimum combination to target response variables and minimize variation

The experimental array selected for this investigation was the Taguchi L_8, with three control factor columns and four interaction columns. The input material was preselected in two categories, thin and thick. (Both groups of input material were within the engineering specification for material thickness.) The experiment was completed with two replicates for thin material and two replicates for thick material. The layout of the experimental array is shown in Figure 16-1. To conserve space, the interaction columns are not shown. The experimental responses, however, were all analyzed for interaction effects.

T R I A L	Control Factors			Response					
	A Blank Length	B ID Tool	C OD Tool	Run 1 thin	Run 2 thick	Run 3 thin	Run 4 thick	Ave. Response	S-N Ratio
1	short	small	small						
2	short	small	large						
3	short	large	small						
4	short	large	large						
5	long	small	small						
6	long	small	large						
7	long	large	small						
8	long	large	large						

Figure 16-1: Experimental Array for Varying Material Thickness

Notice that a signal-to-noise ratio column has been added to the response portion of the array. As we shall examine, the signal-to-noise ratio response is analyzed in the same manner as the average response. The only difference occurs during the selection of levels for the overall optimum response. In the absence of conflicts for multiple responses, the control factor and/or interaction levels, are always chosen to maximize the SNR response. This minimizes variation in the characteristic response in the direction of the chosen target value.

The experimental array for bearing outside diameter with completed response columns is shown in Figure 16-2. Outside diameter is a "nominal-the-best" response that can assume either positive or negative measured values. (Recall from Chapter 13 that the recorded OD values are actually deviations from a target value.) To calculate signal-to-noise ratios we must use the Type II calculation as shown in Equation 16-5.

T R I A L	Control Factors			Response (1=0.0001")					
	A Blank Length	B ID Tool	C OD Tool	Run 1 thin	Run 2 thick	Run 3 thin	Run 4 thick	Ave. Response	S-N Ratio
1	short	small	small	-10.8	-10.2	-9.2	-10.4	-10.2	3.34
2	short	small	large	-9.8	-8.9	-7.0	-8.0	-8.5	-1.30
3	short	large	small	-11.6	-11.3	-12.0	-11.2	-11.5	8.89
4	short	large	large	-10.2	-9.0	-9.3	-9.2	-9.4	5.49
5	long	small	small	-8.5	-9.4	-10.0	-10.3	-9.6	2.01
6	long	small	large	-9.2	-6.7	-7.4	-8.1	-7.9	-0.56
7	long	large	small	-11.5	-10.7	-10.7	-10.8	-10.9	8.26
8	long	large	large	-9.3	-8.8	-8.4	-8.9	-8.9	8.64

Figure 16-2: Experimental Array for Outside Diameter

We should verify, for at least a few trials, that we are able to calculate the Type II SNR and obtain the responses shown above.[9] Now that the SNR's are calculated, let's digress briefly from our analysis efforts to examine their practical significance. The material thicknesses were selected from both the lower and upper bounds of the engineering specification. In actual manufacture, thickness is normally distributed. If we assume this will result in normal distributions for bearing OD (in actuality, it does), we may estimate the distributions of OD for the trials with the lowest and highest signal-to-noise ratios. This experiment indicates a low response of -1.30 dB and a high response of 8.89 dB. These trials are plotted in Figure 16-3 as normal distributions to illustrate the significance of the 10.19 dB range of the experimental results.

---

[9]In this experiment, ten bearings were measured from each trial and run for each of the three output characteristics we are discussing. For the purpose of simplifying this example, we shall treat the responses as if only one measurement was taken for each trial and run.

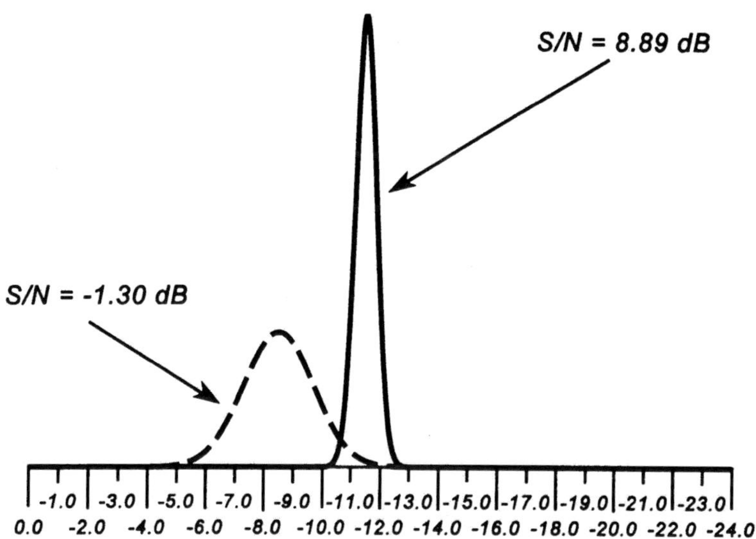

Figure 16-3: Signal-to-Noise Ratios

Since the target value for bearing outside diameter is -12.0, we might examine the responses to our full factorial experimental array and select the combination of forming tools which produces the result closest to -12. We also could look for the trial which demonstrates the least variation. In this instance, both occur with trial number 3, where the average result is -11.5 and the SNR is 8.89 dB as illustrated above. If this were a fractional factorial array the preferred combination would not be so obvious, and we would need to predict the optimum response through the use of the equations presented in Chapter 15.

We might be tempted to stop here, but remember that one of the goals of our experiment was to quantify the effect the control factors and their interactions have on the response characteristics. To do this, we must complete response tables and graphs. The table for OD was previously shown as Figure 15-27. To analyze the signal-to-noise ratio response for OD, we complete a response table as shown in Figure 16-4. We then plot both response graphs, as shown in Figures 16-5 and 16-6.

OUTSIDE DIAMETER SIGNAL-TO-NOISE RATIO															
Trial	Ave.	A1	A2	B1	B2	AB1	AB2	C1	C2	AC1	AC2	BC1	BC2	ABC1	ABC2
1	3.34	3.34		3.34		3.34		3.34		3.34		3.34		3.34	
2	-1.30	-1.30		-1.30		-1.30			-1.30		-1.30		-1.30		-1.30
3	8.89	8.89			8.89		8.89	8.89		8.89			8.89		8.89
4	5.49	5.49			5.49		5.49		5.49		5.49	5.49		5.49	
5	2.01		2.01	2.01			2.01	2.01			2.01	2.01			2.01
6	-0.56		-0.56	-0.56			-0.56		-0.56	-0.56			-0.56	-0.56	
7	8.26		8.26		8.26	8.26		8.26			8.26		8.26	8.26	
8	8.64		8.64		8.64	8.64			8.64	8.64		8.64			8.64
Sum		16.42	18.35	3.49	31.28	18.94	15.83	22.50	12.27	20.31	14.46	19.48	15.29	16.53	18.24
Average		4.105	4.588	0.873	7.820	4.735	3.958	5.625	3.068	5.078	3.615	4.870	3.823	4.133	4.560
Difference		-0.4825		-6.9475		0.7775		2.5575		1.4625		1.0475		-0.4275	

Overall Average Response	4.3463

Figure 16-4: OD Signal-to-Noise Response Table

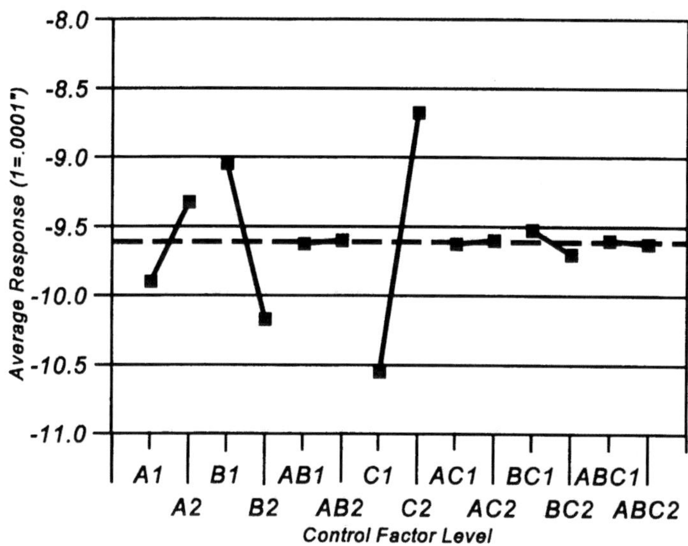

Figure 16-5: Outside Diameter Average Response

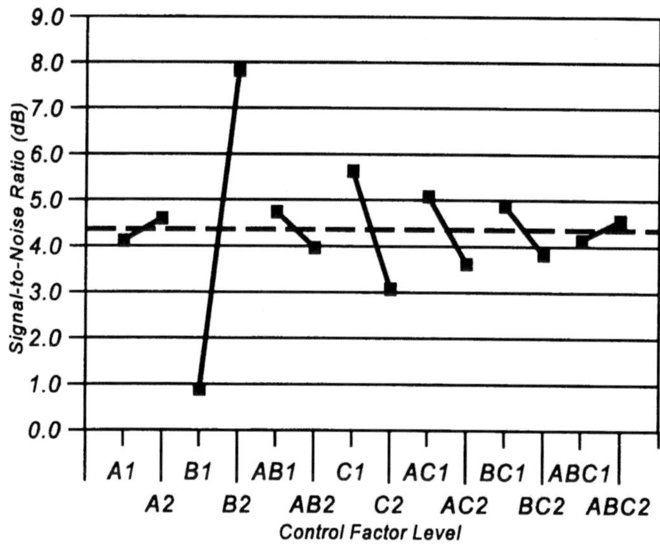

Figure 16-6: Outside Diameter Signal-to-Noise Ratio Response

As we have examined in our earlier analysis of this experiment, the average outside diameter response is controlled by factors A, B, and C. The signal-to-noise ratio response graph indicates that factor B controls the magnitude of variation in the OD response. Specifically, setting factor B at level 2 results in a higher signal-to-noise ratio and less variation in the outside diameter. This is very interesting. We now have the necessary knowledge to control the targeting of the OD response, and minimize variation about the resulting average value.

Obviously, we would choose levels A_1, B_2, and C_1 to control both the targeting and variation in the OD response. This would provide exceptional targeting and variation with respect to the engineering specification of 0 to -24. This response, however, is almost certainly not the optimum response for this process. For instance, our SNR response graph indicates that setting factor B at level 2 results in less variation in the bearing outside diameter. What would happen if we performed another experiment with factor B slightly larger than level 2? Would the OD variation be reduced even more? And, ultimately, how much reduction in OD variation can be achieved? Since factor B also affects the target, we would have to evaluate other levels of factors A and C in order to maintain the process average. To fully optimize this response, sequential experiments must be performed. For illustration, the results of sequential experiments for a different part are shown below (16-7).

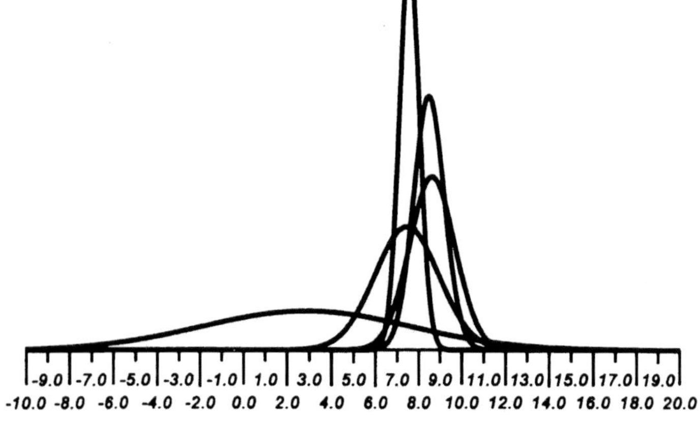

Figure 16-7: Reduced Variation via Sequential Experimentation

Next, we focus our attention on the inside diameter response. The experimental array with the SNR response column is shown below. This response is a positive-only (Type I) characteristic. The SNR is calculated by Equation 16-5. As with the OD array, the interaction columns have been omitted from the illustration, however, they were analyzed in the response table and graph.

T R I A L	Control Factors			Response (1=0.0001")					
	A Blank Length	B ID Tool	C OD Tool	Run 1 thin	Run 2 thick	Run 3 thin	Run 4 thick	Ave. Response	S-N Ratio
1	short	small	small	12.6	11.9	10.0	8.7	10.8	15.63
2	short	small	large	12.6	10.5	8.1	6.7	9.5	11.12
3	short	large	small	16.1	15.9	14.8	14.1	15.2	24.16
4	short	large	large	14.3	13.9	12.8	11.5	13.1	20.38
5	long	small	small	12.0	10.7	8.8	7.3	9.7	13.36
6	long	small	large	12.2	9.1	7.3	6.7	8.8	10.97
7	long	large	small	15.4	15.1	13.3	11.9	13.9	18.58
8	long	large	large	13.9	13.8	12.2	11.1	12.8	19.51

Figure 16-8: Experimental Array for Inside Diameter

The response table for average ID was previously illustrated as Figure 15-28. The response table for inside diameter signal-to-noise ratio is shown on the following page as Figure 16-9. The response graphs for both the average inside diameter and the signal-to-noise ratio are shown in Figures 16-10 and 16-11.

INSIDE DIAMETER SIGNAL-TO-NOISE RATIO															
Trial	Ave.	A1	A2	B1	B2	AB1	AB2	C1	C2	AC1	AC2	BC1	BC2	ABC1	ABC2
1	15.63	15.63		15.63		15.63		15.63		15.63		15.63		15.63	
2	11.12	11.12		11.12		11.12			11.12		11.12		11.12		11.12
3	24.16	24.16			24.16		24.16	24.16		24.16			24.16		24.16
4	20.38	20.38			20.38		20.38		20.38		20.38	20.38		20.38	
5	13.36		13.36	13.36			13.36	13.36			13.36	13.36			13.36
6	10.97		10.97	10.97			10.97		10.97	10.97			10.97	10.97	
7	18.58		18.58		18.58	18.58		18.58			18.58		18.58	18.58	
8	19.51		19.51		19.51	19.51			19.51	19.51		19.51			19.51
Sum		71.29	62.42	51.08	82.63	64.84	68.87	71.73	61.98	70.27	63.44	68.88	64.83	65.56	68.15
Average		17.823	15.605	12.770	20.658	16.210	17.218	17.933	15.495	17.568	15.860	17.220	16.208	16.390	17.038
Difference		2.2175		-7.8875		-1.0075		2.4375		1.7075		1.0125		-0.6475	

Overall Average Response	16.7138

Figure 16-9: ID Signal-to-Noise Response Table

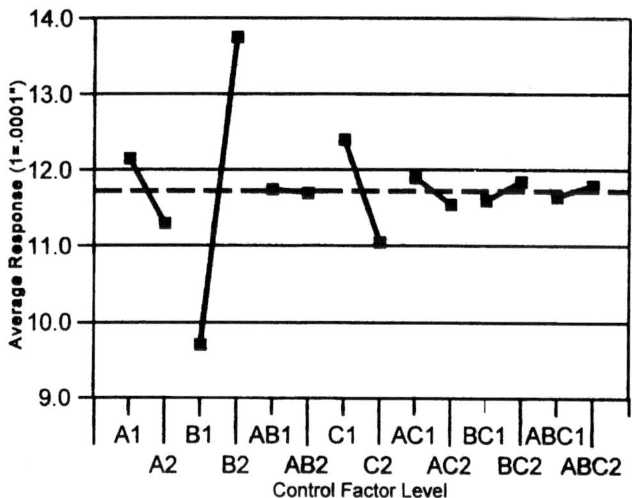

Figure 16-10: Inside Diameter Average Response

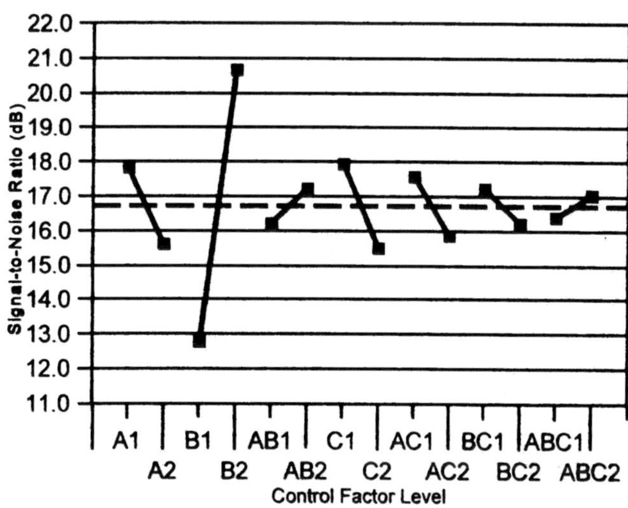

Figure 16-11: Inside Diameter Signal-to-Noise Ratio Response

From the average and SNR response graphs for ID, it is apparent that control factor B has a significant effect on both targeting and variation. As we witnessed with OD, establishing control factor B at level 2 minimizes variation in the output response. The array for the remaining characteristic response, bearing length, is shown in Figure 16-12.

T R I A L	Control Factors			Response (1=1.0000")					
	A Blank Length	B ID Tool	C OD Tool	Run 1 thin	Run 2 thick	Run 3 thin	Run 4 thick	Ave. Response	S-N Ratio
1	short	small	small	.5884	.5885	.5893	.5915	.5894	52.24
2	short	small	large	.5865	.5876	.5881	.5902	.5881	51.57
3	short	large	small	.5890	.5891	.5931	.5929	.5910	48.26
4	short	large	large	.5867	.5889	.5903	.5918	.5894	48.69
5	long	small	small	.5908	.5893	.5919	.5821	.5910	53.26
6	long	small	large	.5897	.5881	.5909	.5913	.5900	52.26
7	long	large	small	.5909	.5906	.5954	.5948	.5929	47.41
8	long	large	large	.5894	.5894	.5938	.5942	.5920	47.89

Figure 16-12: Experimental Array for Length

Like ID, length is a nominal-the-best positive-only response characteristic. The proper equation for calculation of signal-to-noise ratio is Equation 16-5 (Type I). The response table for bearing length has been illustrated in Figure 15-29. The response table for length signal-to-noise ratio is shown in Figure 16-13, on the following page. The response graphs for average length and signal-to-noise ratio are shown in Figures 16-14 and 16-15, respectively.

OVERALL LENGTH SIGNAL-TO-NOISE RATIO

Trial	Ave.	A1	A2	B1	B2	AB1	AB2	C1	C2	AC1	AC2	BC1	BC2	ABC1	ABC2
1	52.24	52.24		52.24		52.24		52.24		52.24		52.24		52.24	
2	51.57	51.57		51.57		51.57			51.57		51.57		51.57		51.57
3	48.26	48.26			48.26		48.26	48.26		48.26			48.26		48.26
4	48.69	48.69			48.69		48.69		48.69		48.69	48.69		48.69	
5	53.26		53.26	53.26			53.26	53.26			53.26	53.26			53.26
6	52.26		52.26	52.26			52.26		52.26	52.26			52.26	52.26	
7	47.41		47.41		47.41	47.41		47.41			47.41		47.41	47.41	
8	47.89		47.89		47.89	47.89			47.89	47.89		47.89			47.89
Sum		200.76	200.82	209.33	192.25	199.11	202.47	201.17	200.41	200.65	200.93	202.08	199.50	200.60	200.98
Average		50.190	50.205	52.333	48.063	49.778	50.618	50.293	50.103	50.163	50.233	50.520	49.875	50.150	50.245
Difference		-0.015		4.27		-0.84		0.19		-0.07		0.645		-0.095	

Overall Average Response	50.1975

Figure 16-13: Length Signal-to-Noise Response Table

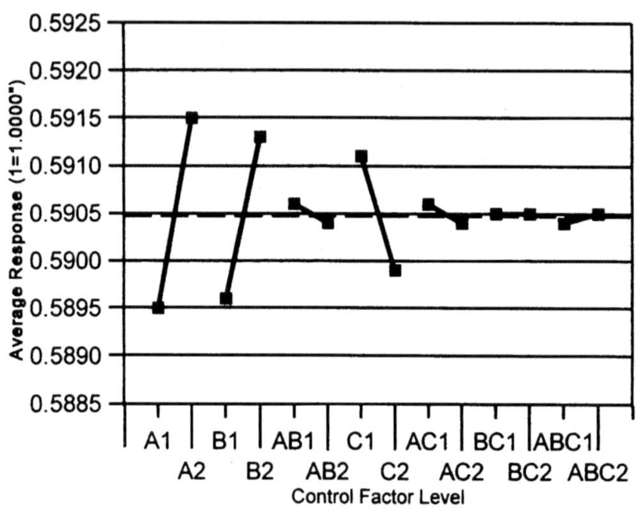

Figure 16-14: Overall Length Response

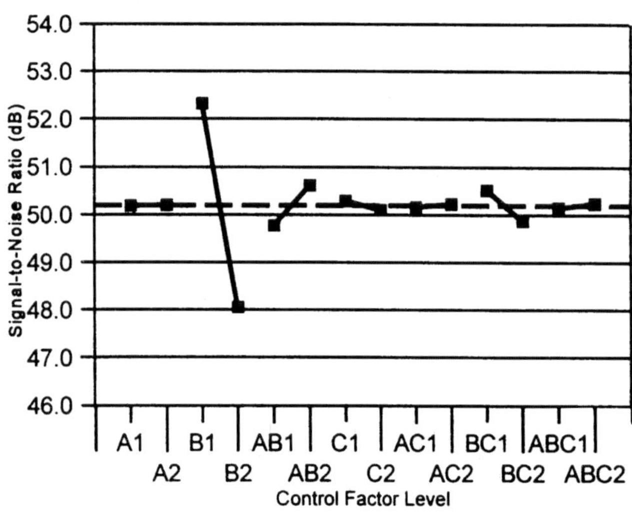

Figure 16-15: Length Signal-to-Noise Ratio Response

The average length response is controlled by factors A, B, and C. The variation in the length response is controlled by factor B, the ID forming tool. This response is opposite of the ones witnessed for OD and ID. In those quality characteristic responses, establishing factor B at level 2 tended to decrease variation in the response. For length, factor B must be set at level 1 to decrease variation. Placing all of these results in a multiple response table yields the following:

Response	Goal	A	B	AB	C	AC	BC	ABC
OD	→ -.0012 ←	A_1	B_2	-	C_1	-	-	-
ID	→ .0015 ←	-	B_2	-	C_1	-	-	-
Length	→ .590 ←	A_2	B_1	-	C_2	-	-	-
OD S/N	↑	-	B_2	-	-	-	-	-
ID S/N	↑	-	B_2	-	-	-	-	-
Length S/N	↑	-	B_1	-	-	-	-	-
Overall Optimum Response		A_1	B_2	-	C_1	-	-	-

Figure 16-16: Multiple Response Table for Averages and Variation

Reviewing the preferred factor levels in each column, it is obvious that we are not able to control targeting and minimize variation in all response characteristics concurrently. We must choose which response characteristics we are going to optimize and which we are going to sacrifice. For this bearing example, OD and ID are the critical characteristics in the customer's application. The levels shown in the last row are the ones which must be chosen to control manufacture of the product.

Inner and Outer Arrays

For the experiment just reviewed, we examined average and signal-to-noise ratio responses when input material thickness was varied between low and high levels. Although material thickness is typically a controllable characteristic, we were treating it like a noise factor within the bounds of its engineering specification. Had the actual experiment consisted of only thin and thick material replicates it would have been, a one-factor-at-a-time evaluation. We stated early in the designed experimentation part of this text that a one-factor-at-a-time approach is not cost efficient, and presents no opportunity to assess interaction effects.

Since thickness variation was placed into only two categories, thin and thick, it is not different than any other two-level factor we have been discussing thus far. It could very easily be placed in a two-level factorial array. That is precisely what was done for the actual bearing experiment. Material thickness was placed in the second column of a Taguchi L_4 experimental array, as illustrated below (Figure 16-17).

Trial	1	2 Thickness	3
1	1	thin	1
2	1	thick	2
3	2	thin	2
4	2	thick	1

Figure 16-17: L_4 Noise Array for Material Thickness

If we turn this L_4 array on its side, the sequence of thin and thick material presentation will correspond with runs 1-4 of the experiment we reviewed. This "sideways" L_4 is shown in Figure 16-18.

3	1	2	2	1
2 Thickness	thin	thick	thin	thick
1	1	1	2	2
Trial	1	2	3	4

Figure 16-18: L_4 Noise Array

This array has two empty rows, for control factors 1 and 3. For the bearing experiment, material source was placed in row 1, with levels of "E" and "C". Machine speed was placed in row 3, with levels of slow and fast. This entire array was then placed immediately after the original L_8 array we have just reviewed, with the L_4 trial numbers corresponding to the "run" columns shown after the L_8 control factor and interaction columns. This is illustrated in Figure 16-19.

This arrangement allows us to evaluate the response characteristics from each trial while the process is being subjected to predetermined and controlled levels of noise. This ensures that the control factor and/or interaction levels selected for the "optimum response" will produce predictable responses across all levels of background noise. We may recall, from Chapter 8, that noise was defined as inputs which affect process outputs but are too difficult or expensive to control. The experimental layout in Figure 16-19 presumes that the factors classified as noise are controllable for a short time period, at least during the course of the experiment.

$L_8 \times L_4$

				source	speed	slow	fast	fast	slow		
					mat.	thin	thick	thin	thick		
					source	E	E	C	C		
TRIAL	A Blank Length	B ID Tool	C OD Tool			Run 1	Run 2	Run 3	Run 4	Ave. Res.	S-N Ratio
1	short	small	small								
2	short	small	large								
3	short	large	small								
4	short	large	large								
5	long	small	small								
6	long	small	large								
7	long	large	small								
8	long	large	large								

Figure 16-19: Inner and Outer Arrays

In the design depicted above, the L_8 array is referred to as the "inner" array, and the L_4 is referred to as the "outer" array. This configuration[10] is one of Dr. Taguchi's prime contributions to process optimization methods. Once factor or interaction levels are chosen to produce the most consistent responses possible when subjected to the noise levels in the outer array, the manufacturing process is classified as *robust*.

---

[10] ©Copyright, American Supplier Institute, Inc., Allen Park, Michigan (U.S.A.) "Reproduced by permission under License No. 970101."

Example 16.2: Optimizing a Laminating Process

In this example, a manufacturer of closures for the food industry was experiencing a problem with a laminated cap liner. The liner consisted of three main components, a paperboard backing, an adhesive layer, and a plastic film as illustrated in Figure 16-20. The plastic film formed the facing (food contact side) of the liner. Its function was to provide an oxygen barrier to prevent spoilage, and provide a food contact surface.

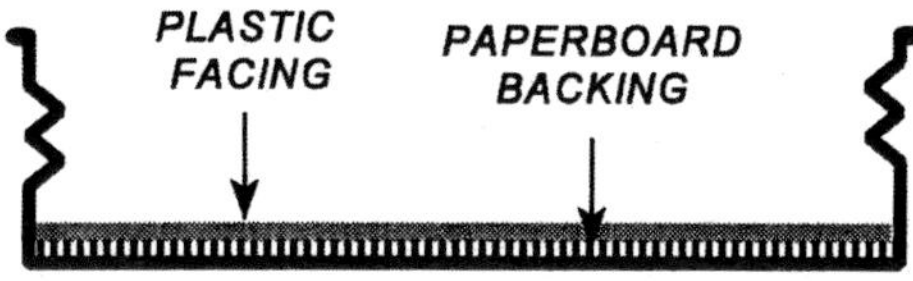

Figure 16-20: Cross-Section of Closure and Liner

Defects occurred when the lamination warped or "curled" after installation inside the metal shell of the closure as shown in Figure 16-21. This defect occurred primarily about one axis and the liner became "u" shaped. The curl reduced the liner's diameter and caused it to fall out of the closure prior to assembly on the container. The rate of returns for affected product lines was in excess of $2,000,000 per year when a designed experiment was initiated to target a curl response of zero.

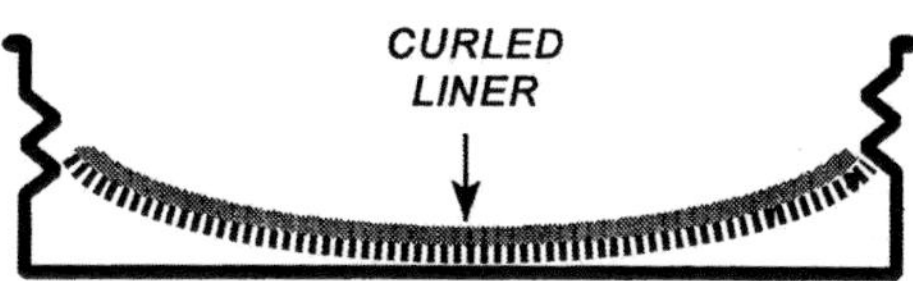

Figure 16-21: Cross-Section of Curled Liner

For the purposes of this experiment, the magnitude of liner curl was determined by placing the liner on a flat surface and pressing an edge in the curled area firmly against the surface. The height of the liner on the opposite edge was measured to quantify the amount of curl.

The curled liner phenomena had been a common, if somewhat intermittent occurrence, for many years. There were a number of theories of root cause for the defect. The first was that tension applied to the plastic film during lamination caused residual stresses in the facing. These residual stresses caused the liner to warp as shown in Figure 16-22. For many years this had led to lamination of these materials at low tension to minimize the theorized residual stresses. Low tension created other difficulties, however, because the facing material had to be taut enough to prevent wrinkles in that layer of the lamination.

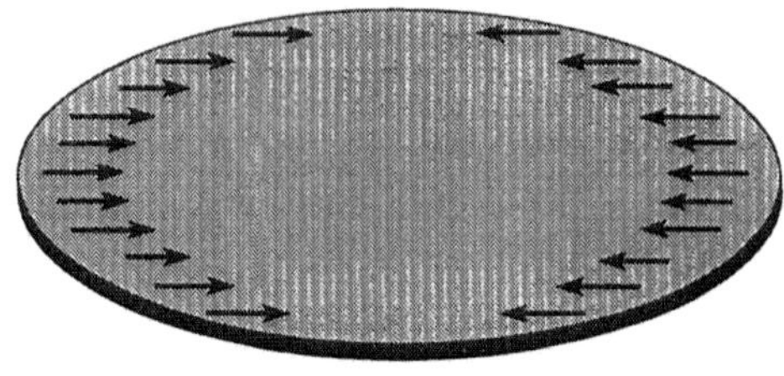

Figure 16-22: Residual Stresses in Plastic Facing

Another failure theory postulated that curling was due to absorption of moisture by the paperboard backing. Specifically, the paperboard would absorb moisture and swell. Since the top face of the liner was a plastic material, the liner would curl around that face as the backing swelled to occupy a larger volume. This, absorption, coupled with the residual stresses discussed above, might cause the liner to curl into the characteristic "u" shape. This is illustrated in Figure 16-23. The action attempted to eliminate this cause was conditioning of materials to control moisture prior to the laminating process and waxing of the paperboard backing. It was hoped these actions would impede moisture absorption.

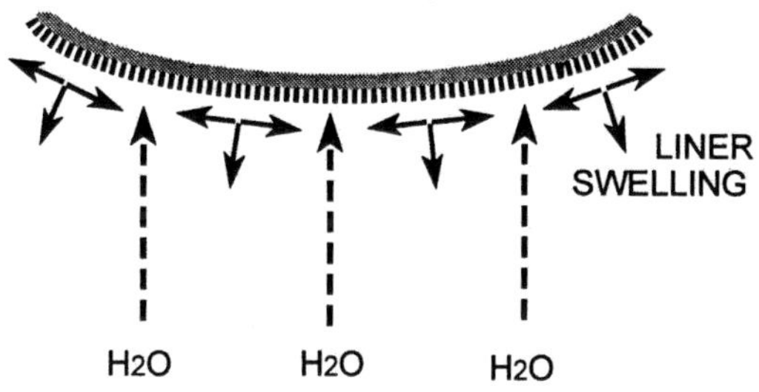

Figure 16-23: Moisture Absorption

Since there were many factors brainstormed as potential causes for this problem, the team formed to resolve the problem chose a fractional factorial design. They decided to use a Taguchi L_8 array and to include noise factors in the experiment. The inputs chosen as control factors are listed in the following table:

Label	Control Factor	Level 1	Level 2
A	Adhesive Type	1	2
B	Backing Type	1	2
C	Supplier	1	2
D	Facing Type	1	2
E	Backing % Moisture	low	high
F	Backing Wax	no	yes
G	Laminator Tension	low	high

Figure 16-24: Laminating Process Control Factors

By definition, noise factors are inputs which are too difficult or too expensive to routinely control. While the definition is somewhat restrictive, it is not uncommon for experimenters to investigate factors based upon their desire to understand the impact of the factor, or based upon a desire to make the process resistant to a specific factor. For this experiment, the experimental team chose the following inputs as noise factors, even though some are controllable:

Label	Noise Factor	Level 1	Level 2
A	Machine	1	2
B	Operator	1	2
C	Roll Position	start	end

Figure 16-25: Laminating Process Noise Factors

The first two factors are somewhat self-explanatory. "Machine" simply referred to the lamination line used to process the paperboard and facing. "Operator" referred to the employee assigned to setup and run the lamination line. The third factor, "roll position" referred to the degree of deformation or set inherent in the paperboard. The paperboard was purchased in large continuous rolls approximately 4' in diameter. Since the outer wraps were from a full roll 4' diameter, they were curved less than the inner wraps which were around the roll's mounting core. The "start" level corresponded to the outer wraps, and the "end" level corresponded to the inner wraps.

The layout of the $L_8 \times L_4$ array is shown in Figure 16-26, on the following page. The completed response table for averages is shown in Figure 16-27.

$L_8 \times L_4$

TRIAL	CONTROL FACTORS								NOISE FACTORS				RESPONSE	
								Roll	start	end	end	start		
								Machine	1	2	1	2		
								Operator	1	1	2	2		
	A Adhesive	B Backing	C Supplier	D Facing	E %	F Backing	G Tension		Run 1	Run 2	Run 3	Run 4	Trial Average	S - N Ratio
1	1	1	1	1	low	no	low		.427	-.163	.280	.273	.204	11.87
2	1	1	1	2	high	yes	high		.222	.043	.173	.078	.129	21.64
3	1	2	2	1	low	yes	high		.172	.097	.128	.134	.133	30.23
4	1	2	2	2	high	no	low		.015	.124	.013	.042	.049	25.67
5	2	1	2	1	high	no	high		.107	.223	.128	.142	.150	25.89
6	2	1	2	2	low	yes	low		.111	.106	.112	.096	.106	42.71
7	2	2	1	1	high	yes	low		.262	.517	.312	.459	.388	18.40
8	2	2	1	2	low	no	high		.230	.287	.233	.247	.249	31.62

Figure 16-26: Experimental Array for Laminating Process

LINER CURL AVERAGE RESPONSE

Trial	Ave.	A Adhesive		B Backing		C Supplier		D Facing		E %Moisture		F Wax		G Tension	
		1	2	1	2	1	2	1	2	low	high	no	yes	low	high
1	.204	0.204		0.204		0.204		0.204		0.204		0.204		0.204	
2	.129	0.129		0.129		0.129			0.129		0.129		0.129		0.129
3	.133	0.133			0.133		0.133	0.133		0.133			0.133		0.133
4	.049	0.049			0.049		0.049		0.049		0.049	0.049		0.049	
5	.150		0.150	0.150			0.150	0.150			0.150	0.150			0.150
6	.106		0.106	0.106			0.106		0.106	0.106			0.106	0.106	
7	.388		0.388		0.388	0.388		0.388			0.388		0.388	0.388	
8	.249		0.249		0.249	0.249			0.249	0.249		0.249			0.249
Sum		0.515	0.893	0.589	0.819	0.970	0.438	0.875	0.533	0.692	0.716	0.652	0.756	0.747	0.661
Average		0.1288	0.2233	0.1473	0.2048	0.2425	0.1095	0.2188	0.1333	0.1730	0.1790	0.1630	0.1890	0.1868	0.1653
Difference		-0.0945		-0.0575		0.1330		0.0855		-0.0060		-0.0260		0.0215	

Overall Average Response	0.1760

Figure 16-27: Laminating Process Average Response Table

The response graph for average liner curl is completed from the data in the response table, and is shown below.

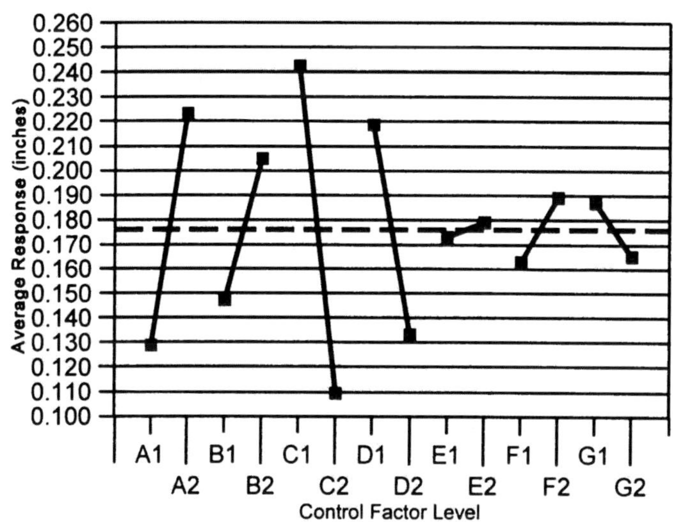

Figure 16-28: Liner Curl Response

A visual scan of the response graph indicates significant control factors are factor C (supplier), factor A (adhesive), factor D (facing), and possibly factor B (backing). Interestingly, the three control factors related to the conventional theories of liner curl (moisture, waxing to prevent moisture absorption, and laminator tension) demonstrated no significant influence on the average curl response.

The response table for the liner curl signal-to-noise ratio response is shown on the following page as Figure 16-29. Since the goal of the experiment was to produce a liner with zero curl, this is a nominal-the-best quality characteristic response which can be either positive or negative. The proper equation for the calculation of SNR is the Type II as shown in Equation 16-7.

LINER CURL SIGNAL-TO-NOISE RATIO

Trial	Ave.	A Adhesive		B Backing		C Supplier		D Facing		E %Moisture		F Wax		G Tension	
		1	2	1	2	1	2	1	2	low	high	no	yes	low	high
1	11.87	11.87		11.87		11.87		11.87		11.87		11.87		11.87	
2	21.64	21.64		21.64		21.64			21.64		21.64		21.64		21.64
3	30.23	30.23			30.23		30.23	30.23		30.23			30.23		30.23
4	25.67	25.67			25.67		25.67		25.67		25.67	25.67		25.67	
5	25.89		25.89	25.89			25.89	25.89			25.89	25.89			25.89
6	42.71		42.71	42.71			42.71		42.71	42.71			42.71	42.71	
7	18.40		18.40		18.40	18.40		18.40			18.40		18.40	18.40	
8	31.62		31.62		31.62	31.62			31.62	31.62		31.62			31.62
Sum		89.41	118.62	102.11	105.92	83.53	124.50	86.39	121.64	116.43	91.60	95.05	112.98	98.65	109.38
Average		22.352	29.655	25.527	26.480	20.882	31.125	21.597	30.410	29.107	22.900	23.762	28.245	24.662	27.345
Difference		-7.303		-0.953		-10.243		-8.813		6.208		-4.483		-2.683	
								Overall Average Response						26.004	

Figure 16-29: Laminating Process Signal-to-Noise Response Table

The response graph for liner curl signal-to-noise ratio is completed from the data in the response table, and is shown below.

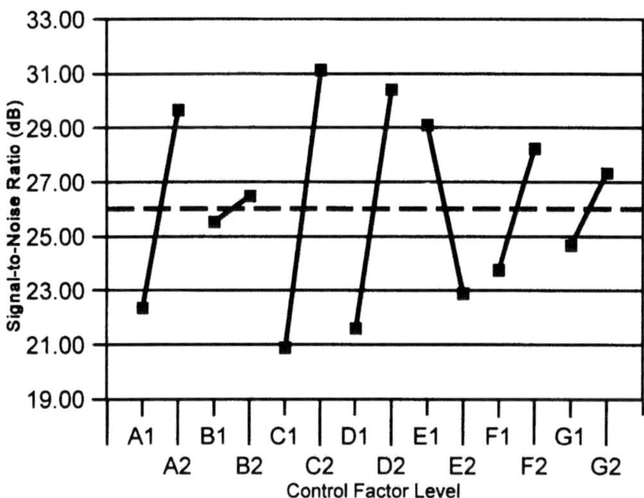

Figure 16-30: Liner Curl Signal-to-Noise Ratio Response

A visual scan of the signal-to-noise ratio response graph indicates factor B (backing type) is the least significant control factor with respect to variation in the liner curl response. The other control factors all produce a significant effect on variation. Even factor G, which seems to graph as small effect relative to A, C, D, E, and F, causes an average change in the SNR response of 2.7 decibels. For nominal-the-best response characteristics, every 3 dB reduction in SNR corresponds to a 50% reduction in the variance (σ^2) of the response.

Returning to the average response graph for liner curl, the manufacturer had to select optimum levels for control factors which affected the curl response. Recall that the target for this response is zero. The control factor levels selected are illustrated in Figure 16-31.

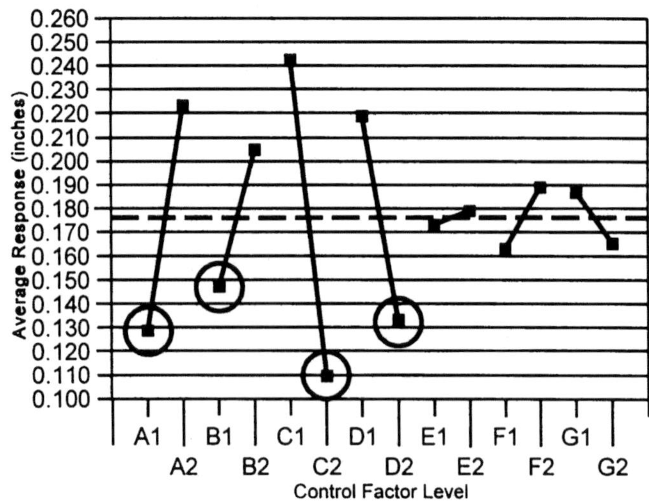

Figure 16-31: Selected Levels to Target Liner Curl at Zero

The manufacturer was uncertain as to whether or not control factor B had a significant influence on the liner curl response. It appeared as though it might, and should therefore be included in the prediction equation. Whether significant or not, the level of control factor B was actually chosen to correspond with paperboard backing type 1, since this was the less expensive of the two types of paperboard evaluated in the experiment. The control factors/levels selected for optimizing the process average response were:

$$A_1 \, B_1 \, C_2 \, D_2 \ \text{ or } \ A_1 \, C_2 \, D_2$$

Assuming that all four of the selected control factors were significant in liner curl reduction, the prediction equation for the overall average response is:

$$\hat{\mu} = \bar{T} + (A_1 - \bar{T}) + (B_1 - \bar{T}) + (C_2 - \bar{T}) + (D_2 - \bar{T}) \qquad (16\text{-}8)$$

$$\hat{\mu} = .176 + (.129 - .176) + (.147 - .176) + (.110 - .176) + (.133 - .176)$$

$$\hat{\mu} = .176 - .047 - .029 - .066 - .043$$

$$\hat{\mu} = -.009 \; \textit{inches}$$

As stated, the manufacturer was uncertain whether or not control factor B was truly significant with respect to the output response. It would have been established a level 1 anyway to minimize manufacturing costs. The prediction equation for the liner curl response, assuming control factor B was not significant, is shown below:

$$\hat{\mu} = \bar{T} + (A_1 - \bar{T}) + (C_2 - \bar{T}) + (D_2 - \bar{T}) \qquad (16\text{-}9)$$

$$\hat{\mu} = .176 + (.129 - .176) + (.110 - .176) + (.133 - .176)$$

$$\hat{\mu} = .176 - .047 - .066 - .043$$

$$\hat{\mu} = .020 \; \textit{inches}$$

The manufacturer was confident if either of these average responses were achieved, it would represent a significant reduction in liner curl. Although not previously discussed, the actual process was being run with control factors established at the same levels as Trial 7 from the experimental array. These settings produced an average response of 0.388 inches, with a signal-to-noise ratio response of 18.40 (Figure 16-26). These results were very close to the data gathered from actual production runs of laminated material.

The manufacturer also had to consider the amount of variation in the liner curl response.   Ideally they would have selected control factors/levels to minimize variation in the characteristic response as much as possible.   This levels are indicated by the high signal-to-noise ratio responses circled in Figure 16-32.

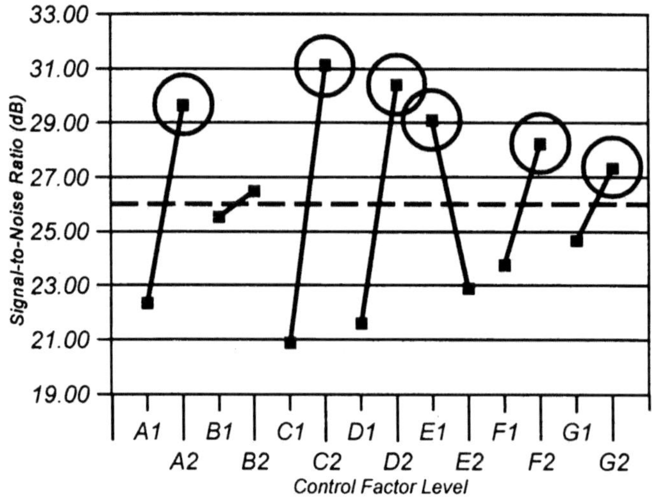

Figure 16-32: Control Factor Levels to Minimize Variation

It was immediately obvious that some of the levels selected to maximize the SNR response (and minimize variation) were not preferable for the process. For example, control factor A was established earlier at level 1 to target the output closer to zero. From the SNR graph above, this level would increase the variation in the response. Also, control factors E, F, and G had indicated no significant effect on the average liner curl response, but were important for minimizing variation. The levels for these factors to minimize variation, however, all added cost to the manufactured product. If possible, the manufacturer preferred to establish these factors at levels to minimize product cost.

A multiple response table was constructed with rows corresponding to each of the manufacturer's desires. The first row indicated the control factors and levels needed to target an average output response close to zero. The second row indicates the control factors and levels to minimize variation. The third indicates the control factors and levels to minimize overall product cost.

Response Category	Goal	Control Factors						
		A	B	C	D	E	F	G
Average	$\rightarrow 0 \leftarrow$	A_1	-	C_2	D_2	-	-	-
Variation	$\downarrow$	A_2	-	C_2	D_2	E_1	F_2	G_2
Cost	$\downarrow$	-	B_1	C_1	D_1	E_2	F_1	G_1
Overall Optimum Response		A_1	B_1	C_2	D_2	E_2	F_1	G_1

Figure 16-33: Multiple Response Analysis by Category

The overall optimum response chosen by the manufacturer is listed above. The predicted signal-to-noise ratio response calculation is shown on the following page. The optimum response chosen by the manufacturer had two additional hurdles to clear before the levels selected became standard operating procedure. A confirmation run needed to be performed, and the levels selected had to minimize liner curl to an extent that no failures would occur and be returned by the customer.

The predicted signal-to-noise ratio response for the levels chosen as the manufacturing optimum is shown in Equation 16-10, below.

$$\hat{\mu} = \overline{T} + (A_1 - \overline{T}) + (C_2 - \overline{T}) + (D_2 - \overline{T}) + (E_2 - \overline{T}) + (F_1 - \overline{T}) + (G_1 - \overline{T})$$

$$\hat{\mu} = 26.00 + (22.35 - 26.00) + (31.13 - 26.00) + (30.41 - 26.00)$$

$$+ (22.90 - 26.00) + (23.76 - 26.00) + (24.66 - 26.00) \qquad (16\text{-}10)$$

$$\hat{\mu} = 26.00 - 3.65 + 5.13 + 4.41 - 3.10 - 2.24 - 1.34$$

$$\hat{\mu} = 25.21 \; dB$$

A series of confirmation runs were performed with the control factors established at the levels shown in Figure 16-33. A series of four runs was needed to correspond to each of the noise conditions found in the outer array for this experiment. The following responses were obtained:

Confirmation Runs	Run 1	Run 2	Run 3	Run 4	Average Response	S-N Ratio
	.014	.152	-.016	.068	.055	22.65

Figure 16-34: Confirmation Run Results

Since the average response in the confirmation run was closer to 0.020" than it was to -0.009", the manufacturer assumed that control factor B was probably not significant with respect to the determination of liner curl. The confirmation run was successful, that is, it confirmed the validity of the experimental results. (With statistical analysis techniques and/or computer programs for experimental design, we can determine confidence intervals around the predicted result.)

The normal distributions representing both the distribution of average responses for the confirmation runs, and the average responses for the normal production runs (Trial 7) are shown below.

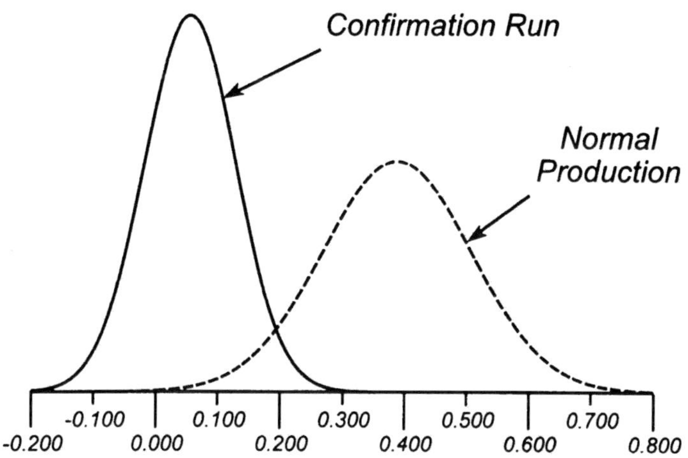

Figure 16-35: Liner Curl Distributions

After the experiment, the standard operating procedures for the manufacturing process were established to run with control factors at the confirmation run levels. This provided better targeting, reduced variation, and the lowest overall manufacturing costs. After implementation, customer returns for this defect category, which were occurring at a rate of $2,000,000/year, ceased to exist.

This example illustrates how to minimize variation, but it also shows that the overall optimum response is influenced by considerations such as manufacturing cost. At times, minimal variation may obtained which exceeds a customer's expectations. This level of performance only has value if it is visible to the customer. When it is not visible, it is best to make a product that meets the customer's expectations at minimal cost. Customers will always prefer the high-quality, low-cost supplier.

References

1. American Supplier Institute *Taguchi Methods, Introduction to Quality Engineering Course Manual,* Version 2.1, American Supplier Institute, Dearborn, MI, 1991.

2. Barker, T. B. *Engineering Quality by Design—Interpreting the Taguchi Approach*, Marcel Dekker, New York, 1990.

3. Barker, T. B. *Quality by Experimental Design*, 2nd edition, Marcel Dekker, New York, 1994.

4. Berger, R.W (ed.) and Pyzdek, T. (ed.) *Quality Engineering Handbook*, ASQC Quality Press, Milwaukee, WI, and Marcel Dekker, New York, 1992.

5. Lochner, R.H. and Matar, J.E. *Designing for Quality—An Introduction to the Best of Taguchi and Western Methods of Statistical Experimental Design*, Quality Resources, New York and ASQC Quality Press, 1990.

6. Montgomery, D.C. and Myers, R.H. *Response Surface Methodology —Process and Product Optimization Using Designed Experiments,* John Wiley & Sons, New York, 1995.

7. Taguchi, G. *System of Experimental Design,* UNIPUB/Kraus International Publications, White Plains, NY, and American Supplier Institute, Dearborn, MI, 1987.

8. Taguchi, G. *Introduction to Quality Engineering*, Asian Productivity Organization, Tokyo, 1986.

17

Implementing Experiments

This text has covered a variety of topics for improving process designs, basic problem solving, and designed experimentation. The most difficult aspect of process improvement in many facilities is obtaining management commitment for improvement projects. All of the examples in this text were initiated and implemented via the rather basic steps that follow. These steps can be used, as appropriate, for simple problem solving and for the initiation of designed experimentation. They are presented as general guidelines for implementation of designed experimentation in a typical manufacturing organization. The actual approach is determined by both the nature of the improvement effort and the operating environment, and is left to the reader's discretion.

Define Experiment Goals

The first step for successful experimentation is a clear understanding of the goal, or goals, of the experiment. Each goal must be a specific, quantifiable entity or quality characteristic. For instance, your desire might be scrap reduction. Your goal would be more specific, such as scrap reduction by elimination of under-length parts. Choosing a specific quality characteristic and action will provide a focal point for the personnel working on the project. It will simplify the brainstorming process, and it will allow the establishment of specific measures to assess the success of your experiment. It is acceptable to select multiple quality characteristics and goals, provided each is clearly defined.

Estimate Experiment Returns

In a typical organization, the next step is to "sell" the experimental approach to management. To accomplish this, one must have some estimate of the potential return of the project. Few organizations can afford the luxury of full scale off-line equipment and facilities for product development efforts. If your organization is like most, experimental trials will be performed on production equipment. This costs money, both in terms of lost production and in terms of lost material and labor resulting from unusable experimental product.

Estimating potential returns at this point is a little vague, however, it is a necessary exercise. Since you have not selected control factors and designed an experimental array, the best you can usually do is assume you will use an array such as an L_8 or L_{12}, and formulate a rough estimate of experimental cost in terms of production units, and production time. (Make certain you allow for both replication of the experiment and for a confirmation run.) Compare this cost to the potential savings of the anticipated improvement.

Even though your estimate will be vague, it should be considered before the experimental effort begins. In most U.S. companies, the costs associated with an improvement effort should not exceed the annual cost savings anticipated. Present the designed experimentation approach to the appropriate levels of management to gain commitment of resources, time, and material.

Select Experiment Team

Manufacturing experiments should not be pursued by individuals. The collective knowledge of a group is far more effective than that of any one person. Also, involving other personnel promotes implementation of experimental results. For instance, if operating procedures are revised, the personnel participating in the experimental effort will help ensure the new procedures are followed. These personnel also become vocal and active supporters, promoting the virtues of the designed experimentation approach. This is often beneficial, since personnel who did not participate in the experiment are expected to discard preconceived and sometimes long-held beliefs about process control techniques.

If the experimental goal is problem elimination, it is wise to select individuals close to the problem, but be sure to include "outsiders" as well. Individuals not directly involved tend to provide a fresh approach and frequently test the validity of accepted theories by constantly asking why and how the process functions.

Brainstorm Control Factors

Construct a Cause and Effect Diagram to develop a list of factors which potentially affect the quality characteristic in question. List all of the possible factors, discuss them, and vote for the ones most likely to control the quality characteristic under study. Categorize the factors as either potential control factors or noise factors.

Consider other product quality characteristics which might be affected by the experiment. A designed experiment will rarely affect only one output characteristic of a manufactured product. Be prepared to measure the effect of the experiment on all output characteristics which may be influenced by the control factors selected.

Design the Experiment

With the consensus of all the personnel working on the experimental team, choose an experimental layout. It should be large enough to address the control factors which garnered the most votes in the brainstorming process, yet it should be small enough to minimize disruption to normal plant operations. That is, it must be cost-effective in the minds of management personnel. Develop a schedule for running the experimental trials. Schedule according to the needs of your organization. For instance, in a production environment, do not schedule a large number of experimental trials for the last week of a production month. Allow adequate time for trial set up. Allow time for measurement and analysis of results. Include time for the confirmation run.

Determine the levels for each control factor in your array. If you are attempting to eliminate a problem in a controlled process, you should not need to establish levels beyond the current specification levels of the control factors. If your goal is new product development, you may purposely choose to establish factor levels higher or lower.

Consider the factors identified as potential noise sources. Determine if blocking or a noise array is appropriate for the experiment.

Perform the Experiment

If the experiment is run on production equipment, involve the operators. (It is usually best if operators are on the experimental team.) Explain the nature and the goals of the experiment. Provide a detailed set-up sheet for each trial to minimize potential for errors. Schedule one or more team members to witness the process set-ups and observe all of the experimental trials. Perform trials in random order and replicate the entire experiment.

Often, experimental results will contradict accepted theories of process relationships, especially the results of individual trials. Never allow unexpected experimental results to be discarded and trials repeated unless assignable causes can be identified which influenced the result! The assignable causes, once identified, must be removed from the process.

While running the experimental trials take detailed notes. Record observations apparent in the trial environment which may be relevant to the process such as time, temperature, humidity. Segregate and clearly identify all experimental parts.

Analyze Experiment Results

Measure the experimental product for the quality characteristic you are investigating, and all other quality characteristics which may be influenced by the control factors in the experiment. Use response tables, response graphs, and interaction graphs to document the results of the experiment. Include a signal-to-noise ratio analysis to document the sources of variation in the quality characteristics. Summarize the results in a multiple response table to provide an overview of the results and select control factor levels to optimize the process. (At times, it may not be possible to optimize all output quality characteristics concurrently. The multiple response table is ideal for documenting the selection of control factor levels.)

Perform Confirmation Run

If the experiment is a fractional factorial, use the prediction equations to estimate the outcome of the optimum combination of control factors. Perform a confirmation run to validate the experiment.

If the experiment is a full factorial, all possible control factor combinations have already been run. In most production facilities, however, it is still advisable to perform a confirmation run at the conclusion of the experimental analysis. This provides validation in an atmosphere removed from the experiment.

In some organizations, it may be advisable to perform two confirmation runs—one to "turn" the improvement on, and another to "turn" the improvement off. This will provide additional assurance that the experimental results are valid. As organizations become accustomed to designed experimentation, confidence in the validity of the approach will increase and the necessity for both positive and negative confirmation runs will be eliminated.

Document and Present Results

The results of the experiment should be documented for future reference. You will also need to prepare at least one summary presentation. At a minimum, this presentation should be given to the experimental team. Members of plant management, engineering, and operators will also benefit by attending a summary presentation of experimental results. If the results of the experiment lead to revisions of standard operating procedures or manufacturing methods, including operating personnel in the summary presentation is crucial to enhance their knowledge of the relationships between control factor levels and quality characteristic outputs.

Implement Results

Actual implementation of experimental results is dependent upon the nature of the control factors and levels selected. Revise standard operating procedures, material purchase specifications or other process controls, as appropriate. Conduct training classes and review the results of the experiment with all affected personnel.

Monitor and Verify Results

After implementation, it is advisable to verify the experimental gains remain intact over time. This may be accomplished through simple monitoring of pre-established reporting mechanisms such as statistical process control charts, quality cost reports or manufacturing scrap summaries. If such reports are unavailable, the process may be periodically audited. This will verify the revised operating procedures and appropriate specifications are being followed.

Appendix

Experimental Arrays and Response Tables

Taguchi L$_4$ Design (2^3)

Trial	1	2	3
1	1	1	1
2	1	2	2
3	2	1	2
4	2	2	1
	a	b	ab

Taguchi L$_4$ Array

Source: ©Copyright, American Supplier Institute, Inc., Allen Park, Michigan (U.S.A.) "Reproduced by permission under License No. 970101."

Response Table for Taguchi L_4

Trial	Response	a		b		ab	
		1	2	1	2	1	2
1							
2							
3							
4							
Sum							
Average							
Difference							

Response Table for Taguchi L_4 Array

Source: Reprinted from *Designing for Quality: An Introduction to the Best of Taguchi and Western Methods of Statistical Experimental Design*, Copyright ©1990 by Robert H. Lochner and Joseph E. Matar, with permission of the publisher, Quality Resources, New York, NY.

Taguchi L_8 Design (2^7)

Trial	1	2	3	4	5	6	7
1	1	1	1	1	1	1	1
2	1	1	1	2	2	2	2
3	1	2	2	1	1	2	2
4	1	2	2	2	2	1	1
5	2	1	2	1	2	1	2
6	2	1	2	2	1	2	1
7	2	2	1	1	2	2	1
8	2	2	1	2	1	1	2
	a	b	ab	c	ac	bc	abc

Taguchi L_8 Array

Source: ©Copyright, American Supplier Institute, Inc., Allen Park, Michigan (U.S.A.)
"Reproduced by permission under License No. 970101."

Response Table for Taguchi L₈

Trial	Res.	a		b		ab		c		ac		bc		abc	
		1	2	1	2	1	2	1	2	1	2	1	2	1	2
1															
2															
3															
4															
5															
6															
7															
8															
Sum															
Ave.															
Difference															

Response Table for Taguchi L₈ Array

Source: Reprinted from *Designing for Quality: An Introduction to the Best of Taguchi and Western Methods of Statistical Experimental Design*, Copyright ©1990 by Robert H. Lochner and Joseph E. Matar, with permission of the publisher, Quality Resources, New York, NY.

Taguchi L$_{12}$ Design (2^{11})

Trial	1	2	3	4	5	6	7	8	9	10	11
1	1	1	1	1	1	1	1	1	1	1	1
2	1	1	1	1	1	2	2	2	2	2	2
3	1	1	2	2	2	1	1	1	2	2	2
4	1	2	1	2	2	1	2	2	1	1	2
5	1	2	2	1	2	2	1	2	1	2	1
6	1	2	2	2	1	2	2	1	2	1	1
7	2	1	2	2	1	1	2	2	1	2	1
8	2	1	2	1	2	2	2	1	1	1	2
9	2	1	1	2	2	2	1	2	2	1	1
10	2	2	2	1	1	1	1	2	2	1	2
11	2	2	1	2	1	2	1	1	1	2	2
12	2	2	1	1	2	1	2	1	2	2	1

This array is not designed for the analysis of interactions.

Taguchi L$_{12}$ Array for Main Effects Only

Source: ©Copyright, American Supplier Institute, Inc., Allen Park, Michigan (U.S.A.)
"Reproduced by permission under License No. 970101."

Response Table for Taguchi L_{12}

Trial	Res.	a		b		c		d		e		f		g		h		i		j		k	
		1	2	1	2	1	2	1	2	1	2	1	2	1	2	1	2	1	2	1	2	1	2
1																							
2																							
3																							
4																							
5																							
6																							
7																							
8																							
9																							
10																							
11																							
12																							
Sum																							
Ave.																							
Difference																							

Response Table for Taguchi L_{12} Array

Source: Reprinted from *Designing for Quality: An Introduction to the Best of Taguchi and Western Methods of Statistical Experimental Design*, Copyright ©1990 by Robert H. Lochner and Joseph E. Matar, with permission of the publisher, Quality Resources, New York, NY.

Taguchi L$_{16}$ Design (2^{15})

Trial	1	2	3	4	5	6	7	8	9	10	11	12	13	14	15
1	1	1	1	1	1	1	1	1	1	1	1	1	1	1	1
2	1	1	1	1	1	1	1	2	2	2	2	2	2	2	2
3	1	1	1	2	2	2	2	1	1	1	1	2	2	2	2
4	1	1	1	2	2	2	2	2	2	2	2	1	1	1	1
5	1	2	2	1	1	2	2	1	1	2	2	1	1	2	2
6	1	2	2	1	1	2	2	2	2	1	1	2	2	1	1
7	1	2	2	2	2	1	1	1	1	2	2	2	2	1	1
8	1	2	2	2	2	1	1	2	2	1	1	1	1	2	2
9	2	1	2	1	2	1	2	1	2	1	2	1	2	1	2
10	2	1	2	1	2	1	2	2	1	2	1	2	1	2	1
11	2	1	2	2	1	2	1	1	2	1	2	2	1	2	1
12	2	1	2	2	1	2	1	2	1	2	1	1	2	1	2
13	2	2	1	1	2	2	1	1	2	2	1	1	2	2	1
14	2	2	1	1	2	2	1	2	1	1	2	2	1	1	2
15	2	2	1	2	1	1	2	1	2	2	1	2	1	1	2
16	2	2	1	2	1	1	2	2	1	1	2	1	2	2	1
	a	b	ab	c	ac	bc	abc	d	ad	bd	abd	cd	acd	bcd	abcd

Taguchi L$_{16}$ Array

Source: ©Copyright, American Supplier Institute, Inc., Allen Park, Michigan (U.S.A.)
"Reproduced by permission under License No. 970101."

Response Table for Taguchi L_{16}

Trial	Res	a		b		ab		c		ac		bc		abc		d		ad		bd		abd		cd		acd		bcd		abcd	
		1	2	1	2	1	2	1	2	1	2	1	2	1	2	1	2	1	2	1	2	1	2	1	2	1	2	1	2	1	2
1																															
2																															
3																															
4																															
5																															
6																															
7																															
8																															
9																															
10																															
11																															
12																															
13																															
14																															
15																															
16																															
Sum																															
Ave.																															
Difference																															

Response Table for Taguchi L_{16} Array

Source: Reprinted from *Designing for Quality: An Introduction to the Best of Taguchi and Western Methods of Statistical Experimental Design*, Copyright ©1990 by Robert H. Lochner and Joseph E. Matar, with permission of the publisher, Quality Resources, New York, NY.

2^2 Western Array

tc	A	B	AB
(1)	-	-	+
a	+	-	-
b	-	+	-
ab	+	+	+

2^2 Western Array

Response Table for 2^2 Array

tc	Response	A		B		AB	
		-	+	-	+	-	+
(1)							
a							
b							
ab							
Sum							
Average							
Difference							

Response Table for 2^2 Western Array

Source: Reprinted from *Designing for Quality: An Introduction to the Best of Taguchi and Western Methods of Statistical Experimental Design*, Copyright ©1990 by Robert H. Lochner and Joseph E. Matar, with permission of the publisher, Quality Resources, New York, NY.

2³ Western Array

tc	A	B	C	AB	AC	BC	ABC
(1)	-	-	-	+	+	+	-
a	+	-	-	-	-	+	+
b	-	+	-	-	+	-	+
ab	+	+	-	+	-	-	-
c	-	-	+	+	-	-	+
ac	+	-	+	-	+	-	-
bc	-	+	+	-	-	+	-
abc	+	+	+	+	+	+	+

2³ Western Array

Response Table for 2^3 Array

tc	Res.	A		B		C		AB		AC		BC		ABC	
		-	+	-	+	-	+	-	+	-	+	-	+	-	+
(1)															
a															
b															
ab															
c															
ac															
bc															
abc															
Sum															
Ave.															
Difference															

Response Table for Western 2^3 Array

Source: Reprinted from *Designing for Quality: An Introduction to the Best of Taguchi and Western Methods of Statistical Experimental Design*, Copyright ©1990 by Robert H. Lochner and Joseph E. Matar, with permission of the publisher, Quality Resources, New York, NY.

Index